STUDIES IN THE HISTORY OF SCIENCE

General Editor: L. Pearce Williams

THE TEMPERATURE OF HISTORY

THE TEMPERATURE OF HISTORY

Phases of Science and Culture in the Nineteenth Century

by STEPHEN G. BRUSH

University of Maryland,
College Park

BURT FRANKLIN & CO., INC.
NEW YORK

© 1978 Burt Franklin & Co., Inc.
New York

All rights reserved.
No part of this book may be reproduced
in whole or in part by any means,
including any photographic, mechanical,
or electrical reproduction, recording, or information
storage and retrieval systems,
without the prior written consent of the publisher,
except for brief quotations for
the purposes of review.

Library of Congress Cataloging in Publication Data
Brush, Stephen G
The temperature of history.
(Studies in the history of science; 4)
Bibliography: p. 135
Includes index.
1. Physics—History. 2. Romanticism. 3. Realism. I. Title.
QC7.B78   530'.09'034   77-11999
ISBN 0-89102-073-X

to Phyllis

Contents

Acknowledgments

I. INTRODUCTION 1
Horizontal and vertical history . . . Movements in science and culture . . . Concepts in the theory of heat

II. ROMANTICISM AND REALISM 15
Origins of "Romanticism" . . . Romantic science . . . The rise of realism . . . Science as part of culture

III. THE AGE OF THE EARTH 29
The second law of thermodynamics and the principle of dissipation of energy . . . Fourier and Kelvin on the cooling of the earth . . . The Evolutionists . . . Radioactivity

IV. PLANETARY SCIENCE: FROM UNDERGROUND TO UNDERDOG 45
Modern attitudes toward planetary science . . . Interactions of planetary and "pure" science in the nineteenth century . . . Isolation and decline of planetary science

V. THE HEAT DEATH 61
Spencer's philosophy of evolution and dissolu-

tion . . . Critique of the principle of irreversibility and of Boltzmann's H theorem . . . The eternal return and the recurrence paradox

VI. REALISM AND NEOROMANTICISM   77

The Prayer Test . . . The reaction against materialism . . . Neoromanticism and positivism . . . Randomness . . . Stallo on the kinetic theory

VII. DEGENERATION   103

Morel, Zola, Baudelaire, Nordau . . . Eugenics . . . Prohibition . . . Decline of degeneration theories

VIII. THE END   121

Henry Adams and the thermodynamics of history . . . Spengler's decline . . . Neorealism and quantification

GENERAL BIBLIOGRAPHY   135

CHAPTER BIBLIOGRAPHIES   137

INDEX   199

Acknowledgments

This book is based on research sponsored by the National Science Foundation. Parts of Chapters I, II, III, V, VI, and VII have been revised from an article published in the *Graduate Journal* of the University of Texas, Spring 1967. A preliminary version of Chapter IV was presented at a workshop on "The Place of the Geophysical Sciences in Nineteenth-Century Natural Philosophy" sponsored by the Hunt Foundation in Pittsburgh, March 1974. The first section of Chapter VI, "The Prayer Test," was originally published in *American Scientist,* October 1974, and is reprinted here by permission of Sigma Xi.

THE TEMPERATURE OF HISTORY

Chapter I

Introduction

> Mathematicians by guiding their thoughts always along the same tracks, have converted the field of thought into a kind of railway system, and are apt to neglect cross-country speculations.
> —(James Clerk Maxwell to Herbert Spencer, 1873, in Duncan 1908: 162)

No one would deny that the development of modern science has been a major factor in the recent history of civilization, yet the relation between scientific theories and general culture is rarely given serious consideration. The magical quality of many technical achievements has combined with the obscurity of scientific writing to hide the fact that scientists have used and been influenced by many of the same ideas that are found in philosophy, literature, and the arts. I propose to examine some of those ideas that played an important role in nineteenth-century theoretical physics, and to suggest how they might be related to trends in other areas of science and the humanities.

Ideas in science and culture may be related in several ways. An idea from culture may enter science, where it can stimulate certain lines of theorizing and (perhaps) suggest new experiments and lead to new discoveries. This was what happened with the romantic concept of the unity of all natural forces. Conversely, scientific facts and theories may have a direct influence on those who construct philosophical systems, write novels, or criticize society. Thus the mechanistic materialism of mid-nineteenth-century physics and biology was reflected by "realism" in philosophy and literature, and by "positivism" in the social sciences. A third possibility is that the same notion may appear at about the same time in both science and culture without any apparent causal influence one way or the other. Such was the case with the principle of dissipation of energy in physics, and the corresponding theory of degeneration in biology, both of which flourished in the pessimistic atmosphere of the latter part of the nineteenth century.

An idea will not be equally fruitful in all places where it is applied. It may be developed so successfully in one field that it becomes well known to us as a basis for modern discoveries, whereas in another it leads only to a dead end and is forgotten. Consequently it is necessary to look at some of the more obscure parts of the history of science and culture if we are to understand the interactions between them.

This book deals with a particular group of scientific theories—those dealing with heat and molecular motion—on the one hand, and a definite sequence of cultural movements—romanticism, realism, and neo-romanticism—on the other. It seems that some of the fundamental concepts of the theory of heat are *Leitmotive* of the cultural movements, and that the philosophical viewpoints of these movements have likewise been reflected in the attitudes of scientists toward the nature of heat.

For those who claim that science is part of culture, or is itself a separate culture, I must note that the word "culture" will be used here in the sense it acquired in nineteenth-century England; it includes literature, painting, music, philosophy, and religion, but *not* science. This exclusion was by no means accidental; the defenders of culture, such as Matthew Arnold, complained that science and technology were destroying their culture and should be de-emphasized. Much as I might wish to believe that science *should* be part of culture, it would be anachronistic to assume it is when discussing the nineteenth century. (A modern representative of the nineteenth-century position is F. R. Leavis, who has rejected C. P. Snow's concept of a "scientific" culture distinct from the literary one. The Leavis-Snow debate is a good illustration of the conflict of the romantic and realist viewpoints discussed below.)

There are two essentially different methods of studying and writing history, "vertical" and "horizontal." (Rand [1971], using an analogy with hydrodynamics, calls them "Lagrangian" and "Eulerian.") The first method, which fits the natural inclinations of those not trained as historians but interested in reading or writing history, is to pick out one particular subject and trace its development from the earliest records to modern times. In the history of science, this is often done by starting from a modern theory or discovery and picking out those earlier events that seem to lead up to it. History is thus presented as *progress* toward the present state of knowledge—a cumulative process in which truth is revealed and error eliminated. The ideas of earlier scientists are judged on the basis of whether they represent a step toward our own ideas. In the past decade this attitude toward the history of science has been labeled "the Whig interpretation" by analogy with Herbert Butterfield's description of the school of historians who viewed British political history as progress toward liberal democracy, and professional historians of science, following Butterfield, have denounced it as a source of misconceptions about what happened in previous centuries.

Another kind of vertical history might be called the "Tory interpretation." According to this viewpoint, which one finds expressed by a few modern writers such as Clifford Truesdell, there was one heroic period or "Golden Age" when all major discoveries in a subject were made, and subsequent work has deteriorated in quality; the only way to make further progress is to soak oneself in the classic writings of the founding fathers and try to fill in a few gaps or think of further routine applications of their principles. This attitude was characteristic of renaissance humanism; the discovery and translation of the documents surviving from Greek antiquity were felt to be the only worthwhile activities. It seemed that no contemporary could possibly match the creative genius of Aristotle, Ptolemy, and Euclid, and indeed this opinion was quite reasonable if one ignored the errors of those men and assumed that they had discovered and developed all by themselves the facts and theories recorded in their writings. But eventually scientists overcame this inferiority complex and made the seventeenth century a new "Golden Age."

During the eighteenth and nineteenth centuries Isaac Newton was supposed (by Lagrange, Laplace, and some British scientists) to have discovered all the basic laws of nature, so that there was really nothing left for his successors to do except apply his equations to new phenomena and work out force laws or mechanisms consistent with his general framework. While there was no reason why scientists as brilliant as Newton could not be born in later centuries, their possible range of accomplishment was thought to be severely limited. Truesdell sees this Golden Age culminating not with Newton but with Leonhard Euler, who put Newton's laws into their final mathematical form and showed how to apply them to the description of the behavior of solids and fluids.

In the twentieth century there was another Golden Age of physics, the period beginning with Planck's quantum theory in 1900 and ending around 1930 with the definitive formulation of relativistic quantum mechanics and the theory of the atomic nucleus. Since then, physics has become more expensive, more esoteric, and more dangerous, but there are some who think it has become more dull, with no giants fit to lift the pencils of Einstein, Bohr, Heisenberg, Schrödinger, and Dirac.

The vertical approach to history tends to underestimate the importance of nineteenth-century physical science, seeing it either as a prelude to the twentieth-century revolution or as a tedious working out of Newtonian principles. The horizontal or "contextual" approach, more often advocated by professional historians, is to pick out a particular historical period, limited in time to a century or less, and study all aspects of the civilization of a certain geographical area, insofar as they have any relation to each other—politics, warfare, economics, philosophy, social conditions, and the arts. Although my discussion centers around the development of a limited number of ideas, I favor the horizontal approach insofar as I prefer to use the ideas to learn something about the relations between science and culture in the West in the nineteenth century, rather than to give a complete chronological account of the ideas themselves. For this reason I shall have little to say about the origins of the theory of heat in the seventeenth and eighteenth centuries, or its modern form in the twentieth century, while at the same time I will discuss many things that may at first seem irrelevant to our subject.

There is nevertheless one serious omission in the horizontal approach as described above: it fails to give much guidance in dealing with major changes that take place between one historical period and the next. That problem has come to the forefront of discussion on the historiography of science in recent years as a result of the popularity of Thomas Kuhn's theory of scientific revolutions. According to Kuhn (1962), each period in the development of a scientific discipline is governed by a "paradigm," or to use his more recent terminology (1974), a "disciplinary matrix." The paradigm includes not only the conceptual framework of accepted theories but also a set of approved worked-out problem solutions ("exemplars") used in training students, and a general viewpoint or set of criteria for determining what kinds of problems and solutions are scientifically acceptable. When a new paradigm is adopted as a result of a scientific revolution, it may not be possible to show that the old paradigm is objectively less satisfactory; each paradigm may be best *by its own criteria,* but a later generation may choose a paradigm that fails to explain certain phenomena in a way that would have been satisfactory to an earlier generation. Thus the seventeenth-century mechanical philosophers sacrificed the Aristotelian concepts of natural place and purpose and the twentieth-century physicists gave up the Newtonian concept of mechanistic explanation; in both cases there was a compensating gain in the power of new theories to describe and predict phenomena.

Against Kuhn's theory of discontinuous change from one paradigm to another in the history of science, critics such as Stephen Toulmin have argued that science moves in a continuous process in which new ideas are generated and accepted or rejected by natural selection; the appropriate metaphor is organic *evolution* rather than political *revolution.* Michael Crowe (1967) suggested that one should recognize periods of accelerated change, such as the 1860s, by calling them *transformational* or *formational* rather than revolutionary. Another proposal, developed by Gerald Holton, is that scientists tend to work in accordance with one or more of a collection of basic themes or "themata," which can never be definitely proved or disproved but may come in and out of fashion. Unlike a paradigm, a theme does not have to dominate the entire discipline, and scientists may hold the same theories with different thematic commitments.

My interpretation of nineteenth-century science and culture does not depend on acceptance of any one of these theories, nor is it completely incompatible with any of them. There does not seem to have been anything like a Kuhnian revolution in the *physical* sciences during the nineteenth century, though one might argue that Darwin's theory of evolution by natural selection established a new paradigm for biology. I would agree with Toulmin that the sharpness of change suggested by the term "revolution" is not an appropriate description of the impact of the discoveries of Darwin, Planck, Einstein—that is, as long as one is comparing these events to sudden South American-style revolutions. But, as Crane Brinton and others have argued, the major political revolutions in modern times have been much slower processes—necessarily so since they resulted in fundamental changes throughout society, not merely a change of leadership. In the history of science we recognize this broader meaning of revolution, as in, for example, the title of Rupert Hall's book *The Scientific Revolution 1500–1800*. If that time span is appropriate for the *first* scientific revolution, it is certainly not unreasonable to claim that the *second* one took at least half as long. So one might well talk about a revolution starting around 1800 and ending around 1950 in which the scientific world view underwent a fundamental change.

This book is not, however, about the Second Scientific Revolution, but rather about the themes and movements which dominated nineteenth-century science and culture in four successive periods. I will explore the hypothesis that, superimposed on the long-term evolutionary or revolutionary change that transformed the Newtonian world view into the modern one, there was a cyclic pattern of changes in ideas. This is not a hypothesis that can be rigorously tested, but it may be illuminating to see how far it can be pushed. I do not claim to have originated the hypothesis—on the contrary, several parts of it will be quite familiar to cultural and intellectual historians—but I think it is fair to say that it has not yet become a respectable interpretation of the role of science in nineteenth-century thought.

The nineteenth century was dominated by three major movements in philosophy and the arts, each of which had its counterpart in the natural and social sciences. We use the commonly ac-

Introduction

cepted terminology for the individual components of these movements, but since there are no standard labels for the movements as a whole (except the first one) we have to choose them somewhat arbitrarily.

The first movement is well known under the name of *romanticism* in the arts; we use the term to include what is known as *Naturphilosophie* in science. Romanticism signifies an emphasis on emotion and spirit as opposed to the rationalism of the Enlightenment; on organic wholeness and synthesis as opposed to analysis and reduction to parts; on philosophical idealism instead of realism; on individuality and nationalism as opposed to universality and cosmopolitanism. (A more complete description will be given in the following chapter.) While the influence of romanticism can be seen throughout the entire nineteenth century, it is only during the period from 1800 to about 1835 that this movement ruled the intellectual world.

Romanticism was followed by another movement which I will call *realism*. It included atomism, materialism, mechanism, naturalism, and certain aspects of positivism. Realism is the opposite of romanticism in many respects, but because it was influenced by romanticism it was not simply a return to the eighteenth-century Enlightenment. This movement reached its height around 1870 and declined thereafter; it was followed by reactions in various different directions.

We use the term *neoromanticism* for the collection of theories and tendencies that arose in the last part of the nineteenth century: aestheticism, decadence, empirio-criticism, energetics, idealism, impressionism, mysticism, sensationalism, and symbolism. Neoromanticism differs from romanticism primarily because it has such a negative character; almost the only common feature of the components mentioned is that they represent some kind of reaction against realism. The later movement was never able to attain that unity of thought and feeling which had characterized the former one.

Following the usual convention in history, we take the First World War as marking the end of the nineteenth century. This allows us to take a brief look at a fourth movement (the completion of two cycles, if you like), *neorealism*.

The division of the nineteenth century into three main periods in this essay is by no means original, but reflects the usage of many other writers. Perhaps the earliest explicit use of the scheme was made by A. V. Dicey (1905) who labeled the "three main currents of public opinion" as follows:

1. the period of old Toryism or legislative quiescence, 1800–1830
2. the period of Benthamism or Individualism, 1825–1870
3. the period of collectivism, 1865–1900

Dicey's scheme was extended to general English history by Somervell (1929), though with an emphasis more on political and literary than scientific trends.

Talcott Parsons (1967) describes two basic viewpoints in nineteenth-century social science:

1. utilitarian (Hobbes, Malthus, physicalistic reduction and "philosophical radicalism")
2. idealistic (Kant and Hegel, leading to holism, Gestalt, and historicism)

These two views, he notes, were criticized and synthesized toward the end of the century by Weber and Durkheim.

Bocheński (1961), describing philosophical movements in the nineteenth century, likewise sees the third period not as a reversion to the first, but as a synthesis of the first and second.

It must be admitted that the cyclic historical pattern-making involved in such descriptions of nineteenth-century thought has often been overdone and is now frowned upon by many professional historians. Much of the writing about these movements has been done from a partisan viewpoint, by critics who are trying to prove something about the evils of modern society. For example, the romantic movement used to be attacked as a detrimental influence on both science and democracy—although the stimulus it gave to the arts could hardly be denied—while the social and scientific accomplishments of the realist period were glorified. Recently this tendency has been reversed, and it has been shown that even in physics the romantic spirit did have some beneficial effects.

But Robert Schofield's book on eighteenth-century British natural philosophy, *Mechanism and Materialism* (1970), shows that such movements can be analyzed in a reasonably detached fashion (cf. Gottschalk [1972]); Christopher Hill's study, *Intellectual Orgins of the English Revolution* (1965), suggests that correlations between fashions in science and culture are of some interest to the general historian. Recent research in music perception, according to Bever and Chiarello (1974), "supports the hypothesis that the left hemisphere is dominant for analytic processing and the right hemisphere for holistic processing," thus indicating that the romantic/realist dichotomy may reflect a fundamental pair of categories in the human mind. Martindale (1975) has shown how psychological analysis of artistic creativity can be applied to historical changes in poetry; his techniques may ultimately provide a basis for understanding the succession of cultural movements.

I now turn to the concepts involved in theories of heat. The first is *caloric*, meaning heat regarded as a substance, presumably some kind of imponderable fluid whose total quantity is conserved as it flows from one place of matter to another. Modern science still retains some vestiges of the caloric theory in terms such as "latent heat" which imply, misleadingly, that a substance "contains" a definite amount of heat. Several of the major scientific theories of the early nineteenth century were apparently based on the caloric theory (Lavoisier's theory of chemical elements, Laplace's theory of the velocity of sound, Fourier's theory of heat conduction, even Sadi Carnot's theory of steam engines) though it was later found that they could be reformulated in a way that did not depend on the assumption of a conserved heat substance.

There was considerable interest in *radiant* heat at the beginning of the nineteenth century; experiments by William Herschel, Macedonio Melloni, James Forbes and others showed that it has the same qualitative properties as light, and suggested that both light and heat should be explained by the same kind of theory. Until about 1820, physicists favored the particle theory of light, and thus the thesis of the similarity of heat and light confirmed the notion that heat is composed of particles rather than being a mode of motion. But Augustin Fresnel's success in establishing the wave theory of light turned this inference upside down, and after 1830 it was generally believed that heat, like light, is a form of vibration in a space-filling medium (the "ether"). The "wave theory of heat," though subsequently forgotten, provided a convenient transition to the modern theory of heat as a form of energy.

The transition from the particle to the wave theory of light was a direct result of theoretical and experimental work within science; cultural movements do not seem to have been involved. But the fact that this transition led to the abandonment of the caloric theory of heat, without any new *scientific* evidence being brought against it, must be regarded as an indication of the powerful drive for *unity* in scientific theories, and that desire does have a cultural correlate in romanticism.

The 1840s saw the establishment of a general principle of *conservation of force,* asserting that all the forces of nature—electricity, magnetism, heat, gravity, mechanical work, etc.—are nothing but different forms of one underlying force in the universe. The word "force" as used in this statement of the principle was considered somewhat ambiguous, and was replaced by *energy.* (Yehuda Elkana has argued that the gradual change in meaning of the word "force" was itself an important part of the discovery of conservation of "energy.") The principle asserts that energy may be transformed from one form to another, provided that the total amount of energy remains constant. The *first law of thermodynamics* is a special case of this principle which applies to transformations of heat and mechanical work.

The principle of dissipation of energy asserts that although the total quantity of energy remains constant, its quality or "usefulness" is continually being degraded. The quality of heat, for example, is measured by its degree of concentration, i.e., its temperature; it is only possible to obtain useful mechanical work from heat if there is available a source of heat at a temperature significantly above that of the environment. The vast amount of heat energy in the ocean is practically useless to man because it is all at about the same temperature. According to Carnot's theorem (1824), the maximum amount of heat that can be converted into work in an engine depends only on the temperature of the source relative to its surroundings, and not on the nature of the "working substance" containing the heat. Moreover, the theoretical maximum

Introduction

work can never be obtained by any real engine because heat tends to flow spontaneously and irreversibly from high temperatures to low; this flow represents a loss of work that could have been obtained by the "ideal Carnot engine."

The *second law of thermodynamics* is a generalization of these statements about the efficiency of engines in converting heat to work. We must call attention to a peculiar aspect of this generalization which is of great significance in modern physics: Carnot's theorem places a limitation on the maximum efficiency of heat engines, but appears to contain no reference to the temporal sequence of natural processes, i.e., the distinction between past and future, and indeed any such distinction is alien to Newtonian physics. Yet by introducing the notion of irreversible heat flow to explain why real engines cannot attain the maximum efficiency, thermodynamics makes a statement about the direction of time in our world.

Further progress in the theory of heat depended on introducing some hypothesis about the atomic constitution of matter. This was one of the major areas of progress in the nineteenth century, for in 1800 the atom was a philosophical concept, not a respectable scientific entity.

As Lord Kelvin remarked in 1870,

> The idea of an atom has been so constantly associated with incredible assumptions of infinite strength, absolute rigidity, mystical actions at a distance and indivisibility, that chemists and many other reasonable naturalists of modern times, losing all patience with it, have dismissed it to the realms of metaphysics, and made it smaller than "anything we can conceive" (Kelvin 1870: 551).*

But, as Kelvin himself showed in the same paper, it was now possible to measure, count, and weigh atoms by means of scientific experiments. Theories based on the assumption that atoms are par-

---

*In the same year W. M. Williams wrote that we are today "as far from a knowledge of . . . the absolute weight of the ultimate atom . . . as the wrangling pedants of the middle ages were from the solution of their much vexed question, of how many human souls could stand on the point of a needle" (Williams 1870: 3).

ticles of small but finite size, moving and colliding according to the laws of mechanics, could provide not only reasonable interpretations of known macroscopic phenomena and predictions of new phenomena that could be tested by experiment, but also, it was hoped, consistent values for atomic properties. As we now know, this hope could not be completely realized as long as the "laws of mechanics" were thought to be those of Newton. Nevertheless, atomism proved to be so successful in most applications that its failures could not be explained simply by denying the existence of atoms. Instead, by pushing the application of Newtonian mechanics to its ultimate limits in the atomic realm, the nineteenth-century theorists both established the reality of atoms as physical objects and prepared the way for the modern quantum theory by revealing some of the inadequacies of classical physics.

The *kinetic theory of gases* is a particular version of the general idea that heat is simply the energy of molecular motion. It includes the further assumption that, most of the time, molecules in gases move in straight lines at constant speed until they strike another molecule or the sides of the container; at ordinary pressures the space occupied by the molecules themselves is only a very small part of the entire gas volume. The gas exerts pressure through the enormously frequent impacts of molecules; this pressure is proportional to the absolute temperature of the gas, which in turn is proportional to the average kinetic energy of the molecules.

Both the caloric theory and the kinetic theory may be regarded as examples of the philosophy of *mechanistic materialism* because they deny the existence of any purpose or "organic principle" in nature, but instead reduce complex phenomena to mere "matter and motion." The difference between them is that the caloric theory illustrates the tendency to seek explanations in terms of *matter*, while the kinetic theory exemplifies the search for descriptions in terms of *motion*. Both assert that the whole is nothing more than the sum of its parts. They do not entirely succeed in doing away with *forces* though there is clearly a preference for explanations based on contact action rather than action at a distance.

The kinetic theory, as developed by James Clerk Maxwell and Ludwig Boltzmann, is the first important example of *statistical explanation* in physics. But the term "statistical" is itself profoundly ambiguous. Its original meaning was quite compatible with belief in an underlying determinism: one uses statistical methods merely for convenience, because it would be too difficult to compute with all the positions and velocities of billions of molecules even though they could in principle be specified. So one proceeds *as if* these positions and velocities were *random* variables, subject to certain general regularities. As one becomes accustomed to this procedure and enjoys some success with it, one puts less emphasis on the assumption that molecules move deterministically, and more emphasis on the statistical properties of the system.

It is of course only a coincidence—pure chance, as it were—that Maxwell announced his statistical theory of molecular velocities in the same year (1859) that Darwin published his *Origin of Species* based on the assumption that random variations are the driving force in evolution. Similarly it is a coincidence that Maxwell presented a critical analysis of the kinetic theory at the same meeting of the British Association for the Advancement of Science (at Oxford in 1860) where Darwin's theory was dissected in the famous Huxley-Wilberforce debate.

Entropy is the term introduced by Rudolf Clausius to facilitate his statement of the second law of thermodynamics. When a body such as the sun at temperature T loses an amount of heat H, its entropy loss is defined as H/T. If that same amount of heat is absorbed by another body, say the earth, at temperature t, the gain in entropy is defined as H/t. The statement that heat always tends to flow from high temperatures to low temperatures implies, mathematically: T is greater than t, hence H/T is less than H/t, hence there is a net increase of entropy. One version of the second law is then simply: entropy tends to increase.

Boltzmann's *H theorem* is the deduction from the kinetic theory of gases that a quantity analogous to entropy, which may be interpreted as the *probability* of a certain type of molecular state, tends to increase as a result of collisions. If the H theorem were valid, it would mean that the kinetic theory of gases, based on Newton's laws of mechanics applied to molecular collisions, implies the principle of dissipation of energy and thus a preferred time-direction of natural processes.

The notion of *degeneration* implies the judgment that an organism or a society is getting worse as time goes on. Insofar as deterioration, decay, and dissolution are associated with the dissipation of energy, we may say that degeneration is the cultural counterpart of the second law of thermodynamics.

Along with the belief in degeneration we often find (as consolation) a belief in rebirth and return to a former "golden age"; this is the theory of the *eternal return*. Strangely enough this idea is not only a pagan myth but also a theorem of mechanics, and as such an inevitable consequence of mechanistic materialism.

These ideas about energy and the general tendency of history intermingle with the three major nineteenth-century movements in a rather complicated way. To oversimplify, we may say that the idea of conservation of energy is a product of the physics of the romantic period combined with that of the eighteenth-century Enlightenment; it provides the basis for the realist physics of the middle of the nineteenth century. The associated idea of the stability of physical conditions over long periods of time underlies the uniformitarian geology of Hutton and Lyell, which in turn makes possible the evolutionary biology of the realist period. (The philosophical idea of evolution was nevertheless largely inherited from romanticism and was given an antirealist interpretation by neoromantic philosophers like Bergson.) The principle of dissipation of energy was developed in the realist period, yet it eventually helped to undermine the foundations of realism. By proving (with the aid of a hypothesis now known to be false) that physical conditions on the earth could *not* have remained constant over very long periods of time, Kelvin managed to cast doubt on the foundations of uniformitarian geology and evolutionary biology. The theory of degeneration likewise originated in the realist period but became one of the motifs of the neoromantic movement. Mechanistic materialism, by then despised but still indispensable in many respects, refined the theory of stability and offered the eternal return as a faint hope of salvation from degeneration.

Chapter II

Romanticism and Realism

> The nineteenth century dislike of realism is the rage of Caliban seeing his own face in a glass.
> The nineteenth century dislike of romanticism is the rage of Caliban not seeing his own face in a glass.
> —(Oscar Wilde 1891: xxxiii)

To introduce the romantic period, and to illustrate the usefulness of the horizontal approach to history, we cite a remarkable example given by A. O. Lovejoy (1936). It might seem that landscape-gardening is a subject fairly remote from philosophy, yet at one point, at least, the history of landscape-gardening must be a part of the history of philosophy. The vogue of the "English garden" (originally the "Chinese garden" in England), which spread rapidly in France and Germany after 1730, was the "thin end of the wedge of romanticism," and foreshadowed a change of taste in all the arts and, indeed, a change of taste in universes. According to Lovejoy,

> In one of its aspects that many-sided thing called romanticism may not inaccurately be described as a conviction that the world is an *englischer Garten* on a grand scale. The God of the seventeenth century, like its gardeners, always geometrized; the God of romanticism was one in whose universe things grew wild and without trimming and in all the rich diversity of their natural shapes (Lovejoy 1936: 16).

The origin of the term "romantic" in this connection is of some interest. If we accept Lovejoy's analysis (1948), we can find the meaning of the word by reading Friedrich Schlegel's essays on ancient and modern literature, written in the 1790s. Schlegel at first shared the prevailing spirit of classicism with its admiration for ancient art as opposed to modern, and he set up the antithesis of *die schöne Poesie* (poetry of beauty) and *die interessante Poesie* (poetry of the interesting) in order to sharpen the distinction. The first type of poetry displays beauty as an objective attribute, and in general, aesthetic values are conceived to be of universal validity independent of the individuals who create or perceive them. The principle of classical art is thus one of self-limitation—exclusion of the intrinsically ugly and of anything that may be inconsistent with the unity of the work of art itself. In order to clarify the definition of classical art, Schlegel also describes the opposite type, which rejects limitations and forsakes universal validity in order to express the richness and individuality of life and nature, including the grotesque as well as the beautiful. The foremost example of *interessant* art is found in Shakespeare's works, but the same general characteristics were attributed to medieval and early modern (i.e., late eighteenth-century) literature.

After 1796, however, Schlegel himself was converted to the new doctrine which he had previously described with disapproval, and he now needed a term that would characterize it better than *interessant*. He therefore selected the term *romantische*, which he had previously used in describing medieval and early modern literature, and attached it to the modern tendency which he now wanted to establish as the highest form of art.

The doctrines of romanticism were propagated early in the nineteenth century by the writings of Schiller, Schelling, Fichte, Heine, and others. The work of Goethe seems to have provided the inspiration for much of German romantic literature, though it is doubtful whether Goethe himself should be called a romantic. The influence of Hegel was predominant in philosophy and history, though many of his followers abandoned romanticism. In painting, romanticism was exemplified by Delacroix in France and the pre-Raphaelites in England.

Historians of music frequently consider the whole of the

nineteenth century to be dominated by romanticism, but the most "typical" romantic composers are said to be Berlioz and Schubert. Romantic music is characterized by the cyclic recurrence and metamorphosis of themes; the classical pause between movements is often omitted. Beethoven is sometimes said to represent the transition between the classic and romantic periods.

Romantic influences on English literature are illustrated by works of Coleridge, Wordsworth, and Blake; on French literature, by works of Victor Hugo, De Musset, and Diderot. American transcendentalism was another offshoot of romanticism.

The romantic viewpoint includes a self-conscious interest in the past; history is not a Whiggish tale of continual progress from the primitive culture of preceding centuries up to the exalted stage of enlightenment of modern times, but rather the record of the numerous possible manifestations of the human spirit, which are not necessarily less worthy merely because they occurred long ago. The "gothic revival" was an attempt to emulate the arts of the Middle Ages; it is exemplified by the novels of Walpole and by architecture which can still be seen throughout Europe and America. The Houses of Parliament in London were rebuilt between 1837 and 1857 in the Gothic style.

The relation of romanticism to political and economic views is more complicated, and varies from one country to another. In Germany and England it goes along with antipathy to the excesses of the French revolution, and therefore tends to be somewhat conservative in tone; whereas in America it is associated with great enthusiasm for democracy and egalitarianism. The only safe generalization here is that romanticism encouraged patriotism as opposed to internationalism.

In religion, one finds a revival of interest in traditional forms: the Oxford movement in England, a "Protestant revival" in Germany, and a "Catholic revival" in France. In America, the transcendentalists were most often found just inside or just outside the Unitarian Church.

The nature of the romantic movement in the arts can be seen in the emphasis on the individual and the unique as opposed to the general, freedom from the restrictions of classical rules of form and structure, direct expression of the emotions, antipathy to academic analysis, and an insistence that the whole is greater than its parts because it is pervaded by a spirit that cannot be rationally explained but can only be intuitively felt. It is these features that also characterize the romantic view of science, usually known as *Naturphilosophie*. The philosophy of nature, as formulated by Fichte, Schelling, and others, is directly opposed to the mathematical-empirical tradition of the seventeenth and eighteenth centuries. It was most influential in biology, where it was associated with vitalism. Its role in the history of the physical sciences has been quite obscure, and has only recently been seriously investigated by historians of science.

The similarity of vitalism to the above-mentioned tenets of romanticism is quite evident. An organism is regarded as an entity that has its own peculiar characteristics, which cannot be reduced to the mere interaction of component parts but must be comprehended as a whole.

But, as noted by Shryock (1947), there are other aspects of the romantic influence, such as the abandonment of quantitative methods in medicine. Speculative theories attributing all diseases to a single cause were rampant. As part of the romantic reaction against the French revolution there was a lack of interest in social planning and public health and welfare, which may have been partly responsible for the rise in the death rate in England after 1810. Curiously enough, humanitarian motives in the study and practice of medicine were stronger in the realist than in the romantic period; not only did doctors become more interested in curing diseases than in classifying them, but they began to use gases to prevent pain in operations; this could have been done as early as 1800 but anesthesia did not become widespread until the 1840s.

Utilitarians and social reformers in England belonged to the realist movement, and were attacked by surviving romantics such as Carlyle, who wanted to preserve some kind of feudal order.

Goethe was deeply interested in biology and was credited with some important discoveries by his admirers. His approach was in general quite dissimilar to that of modern biology, for it was dominated by a desire to interpret the structure and development of organisms with reference to philosophical "ideal types" and did not insist on excluding forces between the atoms, though it preferred to describe long-range forces as being mechanically transmitted through some kind of intervening fluid (ether).

Aside from the revival of Boscovichean atomism, which stimulated some of the work of William Rowan Hamilton in mathematical physics (Kargon 1964), the romantic philosophy did not contribute much to the development of theoretical physics before 1850. The major romantic scientists were not much interested in mathematics except for numerological speculations (Snelders 1973), and they seem to have had little appreciation for the contemporary work of French scientists such as Laplace, Poisson, and Fourier, which was to provide the foundation for modern mathematical physics.

In at least one case there was a head-on collision between a romantic scientist and a realist scientist, in which the latter tried unsuccessfully to establish something like the modern mathematical theory of gases based on atomic motion. In 1820 John Herapath, an English scientist with little formal training who had educated himself by reading the French classics of mathematical physics, sent to the Royal Society of London a paper developing the kinetic theory of gases. According to this theory, the pressure of a gas is primarily due to the motions of atoms, which move in straight lines through empty space until they collide (like billiard balls) with each other or with the sides of the gas container. Humphry Davy was president of the Royal Society at the time, and was mainly responsible for the fate of Herapath's paper. Knowing that Davy had previously criticized the "caloric theory" which represented heat as a substance, and had argued instead that heat is related to atomic motion, one might have expected him to welcome a theory which put his own views into quantitative form. But instead he rejected it on the grounds that it was too speculative, and even though Herapath published his theory elsewhere he never managed to convert any major scientists to his viewpoint (Brush 1957, 1963). Yet it was essentially the same kinetic theory of gases, with a few corrections and improvements, that was almost immediately accepted by all physicists when it was proposed around 1850 by Joule, Krönig, Clausius, and Maxwell.

Mathematics, apparently the most autonomous of all realms of thought, resists any direct affiliation with cultural movements. Nevertheless Norbert Wiener (1951) has claimed that the nineteenth-century concern for rigorous *proof* in mathematics, as opposed to the earlier emphasis on *results*, should be identified as "romantic." Evariste Galois, who was killed in a duel over a woman five months before his twenty-first birthday, certainly fits the stereotype of the romantic hero. His theory of groups is romantic in a different sense: it is a search for structure and symmetry in mathematical systems, not opposed but complementary to the analytic mode of traditional mathematics. The revival of interest in geometry in the early nineteenth century and particularly the invention of non-Euclidean geometries are perhaps symptoms of a concern for global synthesis and logical coherence in contrast to the proliferation of wonderful but somewhat insecure discoveries in the eighteenth century.

It will of course be recognized that something as intangible as a cultural movement cannot be assigned a definite initial and final date, although in a few cases one can point to certain events that appeared to mark a turning point. Even then there is a great temptation to overemphasize the importance of such events, because their significance was usually not clear to contemporaries. As Jacques Barzun remarked (1943: 3), "Romanticism is supposed to have died a hundred years ago. The French date its demise with false precision from the failure of Victor Hugo's last produced play in 1853"; yet others consider that it is still a threat to civilization and that "Fascism and National Socialism are romantic movements based on romantic irrationalism." Likewise it is often said that Wöhler's synthesis of urea in 1828 overthrew vitalism in biology by showing that an "organic" compound could be synthesized from "inorganic" substances; yet recent historical research has demonstrated that this was by no means the significance that Wöhler and his contemporaries attached to his work (Lipman 1964). But Everett Mendelsohn (1964, 1965) argues that *Naturphilosophie* was firmly rejected by mid-century biologists even though they still talked about vital forces.

Whatever date one may wish to assign to the death of romanticism, there is no doubt that there *was* a counter-movement in the "spiritual forces." His research on color perception, in itself quite respectable, was unfortunately combined with a polemic against Newton's theory of colors and an attempt to establish a new theory of his own. Encouraged by the applause of Schelling and the other romantic philosophers, Goethe continued his scientific work, which culminated with his theory of the "spiral tendency" of plants. According to this theory, a plant is composed of two opposing tendencies: the vertical, which represents the male, eternal essence, and the spiral, which represents the female, nourishing principle. (This *vive la différence* approach to sex may be contrasted with the feminist philosophy of realists like Mill.)

The romantic viewpoint in science consists in interpreting all phenomena in terms of a single basic principle, or perhaps as the result of two basic contrasting principles. As it happened, such a viewpoint could be quite fruitful at certain stages in the development of science, although at other stages it might be harmful. Biology was not quite ready for it at the beginning of the nineteenth century; there were still too many empirical facts yet to be discovered and classified before it would be possible for a grand generalization to produce anything but confusion. Chemistry could profit by it to some extent, though even there it was dangerously easy to carry speculation too far, and the net result was a relapse into empiricism which discouraged chemical theorizing later in the century. It was primarily in physics that conditions were favorable for a leap of the imagination. By the end of the eighteenth century a great deal was known about light, electricity, magnetism, heat, and gravity, and this knowledge could be described theoretically by an atomic theory in which each atom was surrounded by several imponderable fluids. But the model was becoming clumsy, and it must have occurred to many scientists that there should be a simpler theory which would exhibit all these different phenomena as manifestations of one or two basic principles. The romantics found the answer in the philosophy of Immanuel Kant, which proposed to replace the mechanical picture of atoms moving through empty space by a system of attractive and repulsive forces. These forces fill all space and make the separate existence of matter almost unnecessary. Atoms, if they exist at all, are merely centers of force as in the earlier theory of Boscovich.

German romantic philosophy, based on the "dynamic" physics of Kant as opposed to the mechanist or "materialist" physics of Laplace and other French scientists, was transmitted to England by various means, one of which, according to L. Pearce Williams (1964), was the direct personal influence of Samuel Taylor Coleridge. Coleridge visited Germany in 1798 and became an enthusiastic adherent of Kant's philosophy; when he returned to England he met Humphry Davy, who was beginning to doubt the theory of imponderable fluids and was apparently receptive to the dynamic viewpoint. Davy in turn influenced his pupil Michael Faraday, and in this way the notion of a physical universe dominated by various manifestations of underlying attractive and repulsive forces provided the motivation for some of the most important work in experimental physics in the first part of the nineteenth century. Another example of the romantic influence is the work of Oersted, who sought for twelve years and finally found a connection between electricity and magnetism because Schelling's *Naturphilosophie* gave him the conviction that such a connection must exist (Stauffer 1953, 1957). The doctrine of the essential unity of all forces in nature leads directly to the law of conservation of energy, in which the term "energy" replaces "force" as the label for the quantitative magnitude that remains numerically constant when one kind of force is converted into another. (The particular case of the interconversion of heat and mechanical work is also known as the first law of thermodynamics.)

The idea that the discovery of the conservation of energy owes something to the romantic philosophy is by no means accepted by all historians of science, and it is certainly true that other factors were also involved. The reader who wishes to settle this question for himself is advised to start by reading Robert Mayer's paper of 1842, "Remarks on the Forces of Inorganic Nature," which is generally regarded as being one of the first explicit general statements of the law of conservation of energy. Whereas Mayer refused to regard any of the observed forms of energy as being more fundamental than the others, other scientists soon jumped to the conclusion that all forms of energy could be represented as *mechanical* energy; thus heat could be considered as nothing but the kinetic energy of motion of atoms. This new version of the mechanical philosophy arts, known either as "realism" or "naturalism," and a corresponding counter-movement in the physical and social sciences which has been called "materialism," "positivism," or "mechanism." In literature, we find the attempt to portray life as it really is, with all its sordid and trivial aspects, without idealization. Flaubert's *Madame Bovary* is sometimes said to be the most famous realist novel; other writers associated with the movement were Balzac, Zola, Dostoevsky, Hardy, Dickens, Walt Whitman, and Gogol. Goya, Daumier, and Courbet were realist painters. In architecture, the gothic style was followed by an emphasis on more functional and economical design ("organicism" and "mechanism") and increasing use of metal in construction.

The category of realism is somewhat more obscure in music, but it has been claimed that Modeste Mussorgsky was a realist in music because "he dramatized not himself or his own private emotions, but those of whatever subject he had in hand. This is what gives his music the impersonal aspect that eventually was to guide all music out of the romantic impasse" (Copland 1968: 27). Wagner has been called a realist in a somewhat different sense (Barzun 1941). Later on there was a minor musical movement called *verismo*, exemplified by operatic librettos that presented ordinary people in familiar situations engaging in outbursts of realistic brutality: Mascagni's *Cavalleria Rusticana* (1890) and Leoncavallo's *Pagliacci* (1892).

The major philosophers of the realist period were Comte, Herbert Spencer, J. S. Mill, Feuerbach, and Kierkegaard. One may find some difficulty in perceiving a common denominator in this group, and we shall claim no more than that they all represent in some way a return to rationality and empiricism after the speculative excesses of the romantic period. Political liberalism and the cause of women's rights became respectable. Similarly in religion and theology, it was somewhat less dangerous to advocate unitarianism, agnosticism, or even atheism; the latitudinarian or broad-church movement in England stressed toleration, while the "higher criticism" tried to apply objective standards to the analysis of scripture.

But it was in the social and biological sciences that the realist movement made its greatest impact on intellectual history. I refer primarily to the theory of evolution by natural selection (Darwin and Wallace) and the materialistic theory of history (Marx and Engels), but also to the foundation of psychophysics (Weber, Fechner), the application of statistics to the study of society and of human heredity (Quetelet, Galton), the cell theory in biology (Brown, Schwann, Schleiden), the germ theory of disease (Koch, Pasteur), and the introduction of physical and chemical methods into physiology (Helmholtz, Ludwig, Emil du Bois-Reymond). The valence theory of chemical bonds was developed during this period, and it may perhaps be characterized as a mechanistic theory in contrast to the earlier "dualistic" electrochemical theory of Davy and Berzelius, which attempted to explain all chemical combinations by means of simple attractive and repulsive forces. There were two major achievements in theoretical physics, both of which were based on experimental work of the earlier period. The first was what was then called the "mechanical theory of heat," which included both thermodynamics and the kinetic theory of gases. Thermodynamics could be treated purely as a quantitative description of macroscopic phenomena, and did not necessarily imply an atomistic conception of matter. The kinetic theory definitely did require such a conception, and may be regarded as the most extreme form of mechanistic materialism. The second major achievement was Maxwell's theory of electromagnetism, which developed out of Faraday's lines of force and some of Kelvin's mechanical models, but eventually came to be regarded, like thermodynamics, as a mathematical description of macroscopic phenomena independent of any microscopic or atomistic theory.

In mathematics, Riemann was the leader of a new age of analysis, developing a powerful calculus of real and complex variables; he founded differential geometry, converting what had previously been the study of forms to a more quantitative basis, and discovering by this method another version of non-Euclidean geometry which implied a closed rather than an open universe. Riemann's work, unlike that of many of his predecessors and successors, had a close affinity to physics; it is not too difficult to see it as part of the realist movement in science.

Can science be part of a cultural movement?

The scientific advances of the realist period were so great, by comparison with those of the romantic period, that historians have tended to regard romantic science as either worthless speculation or a collection of unrelated discoveries which properly belong to the later period. Conversely, the influence of romanticism on the arts was so great, by comparison with that of realism, that the latter has seldom been considered as a separate period worthy of study in its own right, but has instead been treated as a temporary reaction against romanticism or as a mere working out of some of its minor aspects (Barzun 1943). In other words, all scientists in the nineteenth century were really realists and all artists were really romantics. But I claim that there was a romantic and a realist period in both the sciences and the arts, and that there was a very definite correlation between the swings of the pendulum in many different areas.

The idea of different styles or viewpoints in the arts, exhibited by different historical periods, is widely accepted and needs no comment here. Likewise in philosophy the first question one asks about any thinker is: realist or idealist? The assignment of philosophers to categories, and the determination of historical periods of realism and idealism in philosophy, are full-time academic activities. But in the history of modern science it has frequently been asserted that the personal philosophical viewpoints of scientists have very little role to play. Wilhelm Ostwald's description of scientists as "classics" or "romantics" touched only on their personalities, not the kinds of theories they would accept. It is only recently that historians of science have begun to analyze the effect of such extrascientific hypotheses or presuppositions about scientific methods and theories on the development of science itself. Of course the supposition that these factors should not influence scientific work has made scientists themselves reluctant to admit their influence on their own discoveries, and it is therefore more difficult to find definite evidence than in the history of literature or painting.

Why should there be a correlation between the movements in the sciences and in the arts? I suggest two major reasons, both of which are less valid today than they were in previous centuries. First, in many instances it was actually the same people who were leaders in several different areas, so that it is plausible that they would exert the same type of influence in all of these areas. Second, there was a considerable amount of communication between different disciplines by means of popular periodicals, lectures, and books. These factors combined to produce a very strong correlation in the romantic movement, as with earlier movements such as the eighteenth-century enlightenment, but after that time the interactions between the arts and sciences were weaker. It may be only accidental that the realist reaction to romanticism occupied about the same period of time in different areas, and was then followed in each case by a reaction against realism. By the time we get to the reaction against the reaction, much of the original coherence has been lost. The cultural tendencies at the end of the nineteenth century, which I have lumped together under the heading of neo-romanticism, contain such divergent elements as spiritualism, symbolism, aestheticism, impressionism, political reaction and monarchism, ultranationalism, and neo-Catholicism; it is perhaps a strain on one's credulity to interpret them as part of the same movement as neo-Kantian idealism, empiriocriticism, energetics, axiomatic mathematics, neovitalism and degenerationism in biology, *Gestalt* psychology, and introspection.

One characteristic of nineteenth century science should be emphasized: it had developed far enough to attract general attention, but it had not yet become so esoteric that the nonspecialist was afraid to stick out his neck to comment on it. Whereas today science influences culture mainly through technology while its intellectual basis is incomprehensible to almost everyone (including many of the scientists themselves!), in the nineteenth-century science was a powerful cultural force as a generator and critic of ideas that were not too difficult to be understood by intelligent educated readers. Thus one finds that the latest discoveries and theories of science were being discussed, in considerable depth and detail, in intellectual reviews that were the equivalent (at least in circulation) of today's *Encounter, Atlantic Monthly,* and *New York Review of Books.* For example, the *Edinburgh Review,* founded in 1802, dealt with science as well as literature, and one of its first numbers carried an attack on Thomas Young's wave theory of light by Henry Brougham, which may have retarded the acceptance of that theory by British scientists. There was a considerable body of scientific amateurs, especially in the biological and geological sciences, who kept up with the progress of science through such journals as the *Philosophical Magazine* and later *Nature,* and by attending meetings of the British Association for the Advancement of Science. A lavishly illustrated book on natural history or even on spectrum analysis was apparently considered "equally adapted to the drawing-room table as an *édition de luxe,* and to the study of the philosopher as a book of reference" (Pritchard 1869: 490).

There were equivalent journals and societies in France and Germany; indeed, it was Lorenz Oken, one of the romantic biologists, who organized the first meetings of the German scientific society, and published a journal of general science. Even the mathematical sciences were not beyond the reach of the layman, for in England a first class honors degree in mathematics from Cambridge or Oxford was considered to be the ideal preparation for a brilliant career in law or the Church.

I will return to the neoromantic period in Chapter VI, but first it is necessary to describe in more detail the development and influence of the theory of dissipation of energy, which served to connect the realist and neoromantic periods, though it undermined the unity of science.

Chapter III

The Age of the Earth

> If there be heat in the centre of the globe, it must have the properties of heat and none other. I ask not how the Heat originally was lodged in that situation, for the origin of all things is obscure; but I ask why, in the countless succession of ages which the Huttonian requires, the Heat has not passed away by conduction, and if it has passed away, by what other heat it has been replaced?
> —(George Greenough 1834: 62)

The first law of thermodynamics (conservation of energy), inspired in part by the philosophy of romanticism, provided an organizing principle for the science of the realist period. Likewise the second law of thermodynamics (dissipation of energy), which arose from the technical analysis of steam engines, provided a *dis*organizing principle which turned out to be highly appropriate for the neoromantic period. This later period was characterized by pessimism about the future of the human race and its democratic forms of social organization. Artists and writers were throwing off the chains of traditional rules of style and taste in their pursuit of aesthetic ideals. Scientists were beginning to renounce the possibility of a rational understanding of man and nature by means of the causal laws and ordered categories of the realists, and were falling back on empiricism and sensationalism. In the midst of this atmosphere of disintegration, the second law of thermodynamics asserted that the entire universe is running down; the amount of energy available for doing useful work is always decreasing, as entropy (representing randomness and disorder) increases relentlessly. This concept of the irreversibility of physical processes forms a startling contrast to the reversibility and regularity of celestial mechanics. The heavenly clockwork which, according to the eighteenth-century mathematicians, needed no God to keep it going once it was started, could work only in the vacuum of outer space; something went wrong as soon as one tried to bring it down to earth and use it to explain terrestrial phenomena. Nature, it seemed, would not maintain indefinitely long the stable environment which would permit the slow but sure action of geological and biological processes in the mechanistic manner postulated by midcentury naturalists. Dissipation made evolution and progress seem doubtful.

As is well known, the "second law of thermodynamics" was discovered by Sadi Carnot (1824) in the course of his studies on the efficiency of steam engines, and the problem of obtaining the maximum amount of work from a given amount of fuel seems to have provided the motivation for many of the nineteenth-century researches on heat and its transformations. Although one can find scattered statements in the literature before 1850 to the effect that something is always lost or dissipated when heat is used to produce mechanical work, it was not until 1852 that William Thomson (later Lord Kelvin) generalized the second law of thermodynamics and asserted the existence of "a universal tendency in nature to the dissipation of mechanical energy." At the same time he mentioned his conclusion that

> Within a finite period of time past, the earth must have been, and within a finite period of time to come the earth must again be, unfit for the habitation of man as at present constituted, unless operations have been, or are to be performed, which are impossible under the laws to which the known operations going on at present in the material world are subject (Kelvin 1852: 306).

I have quoted the whole sentence including the hedge, for Kelvin always wanted it to be clearly understood that his conclusions did not necessarily apply to living matter, and did not exclude the possible intervention of nonphysical agents in the world.

The consequences of Kelvin's dissipation principle were elaborated further by Hermann von Helmholtz in a lecture two years later, in which he described the final state of the universe: all energy will eventually be transformed into heat at uniform temperature, and all natural processes must cease; "the universe from that time forward would be condemned to a state of eternal rest." Thus was made explicit the concept of the "heat death" of the universe.

The modern statement of the dissipation principle involves the notion of "entropy," introduced by Rudolf Clausius in 1865. It should be emphasized that no new physical content was being added to the second law of thermodynamics in this formulation of Clausius, yet the mere act of giving a new short name to a physical quantity that had previously been represented only by mathematical formulae and awkward circumlocutions had an undeniable influence on the subsequent history of thermodynamics. Entropy is taken from the Greek $\eta\tau\rho o\pi\eta$, meaning transformation, and was intentionally chosen to have a resemblance to the word energy. The two laws of thermodynamics are then

1. The energy of the universe is constant.
2. The entropy of the universe tends toward a maximum.

The significance of Kelvin's enunciation of the dissipation principle in 1852 cannot be understood adequately as merely a technical contribution to thermodynamics; instead, one must notice its relation to contemporary developments in geology, and to Kelvin's own previous work on Fourier's theory of heat conduction. In fact, a principle of irreversibility is explicit in Fourier's first formulation of his theory, even though, in accordance with the caloric theory, he did not consider changes in the total *amount* of heat in a system. Fourier wrote in 1807:

> When heat is unequally distributed among the different points [parts] of a solid body, it tends to come to equilibrium and pass successively from hotter to colder parts. At the same time the heat dissipates itself at the surface and loses itself in the surroundings or the vacuum. This tendency toward a uniform distribution, and this spontaneous cooling which takes place at the surface of the body, are the two causes which

change at every instant the temperature of the different points (translated from Fourier 1807: 33).

The importance of Fourier's statement lies in the fact that it is a prelude to the first *quantitative* statement of a general irreversible law of physics, his differential equation for heat flow.

It seems to have been Kelvin's early studies of Fourier's theory, started while he was still an undergraduate at Glasgow, that first set him thinking about the past history of the earth. Fourier's book *Théorie Analytique de la Chaleur* (1822) was destined to become the foundation for a major part of modern mathematical physics, but in 1840, when Professor John Pringle Nichol introduced his bright student William Thomson to it, it was still practically unknown in Britain. Phillip Kelland, Professor at Edinburgh, had stated (1837) that most of Fourier's results were wrong, and Kelvin's first published paper (at the age of seventeen) was devoted to an exposition of Fourier's theory in which he showed that Kelland's criticism was based on his misunderstanding of the fact that certain functions may be expanded in series consisting of only sines or only cosines. Professor Kelland never recovered his reputation in physics after that episode.

At the end of his fourth paper on Fourier's theory, in 1842, Kelvin pointed out that when negative values of the time variable are substituted into the solution of the heat conduction equation for a specified temperature distribution at time $t = 0$, there is in general no meaningful result; in other words, an arbitrary initial distribution cannot in general be produced by some previous possible distribution. Many years later, Kelvin referred to this result as a mathematical deduction that there must have been a creation (see S. P. Thompson 1910). His colleague Peter Tait proclaimed this conclusion in 1871; referring to Kelvin's geological applications of the dissipation principle, he said it had been established that "the present order of things has *not* been evolved through infinite past time by the agency of laws now at work, but must have had a distinctive beginning, a state beyond which we are totally unable to penetrate, a state, in fact, which must have been produced by other than the now acting causes" (Tait 1871: 6).

That view inevitably came into sharp conflict with the "uniformitarian" philosophy of geology, which was replacing the "catastrophist" interpretation of the earth's history as romanticism gave way to realism. According to the uniformitarians, the present appearance of the earth's surface should be explained as a result of physical causes, like erosion, whose operation can still be observed, rather than by postulating catastropic upheavals in the past. The leader of the uniformitarian school at midcentury was Sir Charles Lyell, whose book *Principles of Geology* (1830-33) is still regarded as the first comprehensive modern text on geology. To some extent Lyell was simply reviving the earlier theories of James Hutton, but he was able to draw together a much larger mass of evidence in a convincing way; he seems to have been mainly responsible for finally eliminating the Flood as a respectable component of geological speculation.

The chief characteristic of uniformitarian geology that concerns us here is that (like theories of biological evolution) it requires an immense span of time in which certain causes may operate in order to produce the present configuration of the earth's surface. This was its weak point, according to Kelvin, for his calculations based on heat theory indicated that when proper account is taken of the rate of dissipation of energy, it can be shown that the physical state of the earth and sun (especially the temperature) could not possibly have remained sufficiently constant over such long periods of time.

Rudwick (1974) suggests that uniformitarian geologists thought of time in the same way economists thought of money: you can always issue more of it when needed. Kelvin wanted to return to a world with a fixed amount of temporal currency. He did have good scientific reasons for his belief that things are generally cooling down. Many geologists, in spite of Lyell, thought that the nature of rocks and other formations on the earth's surface furnished ample evidence that temperatures had been much higher in the past. As for the sun, Kelvin himself accepted for a few years the suggestion of J. J. Waterston (1853) that the infall of meteoric matter could provide enough energy to replace that which is radiated away. But he concluded that the meteoric fuel must be assumed to circulate inside the earth's orbit for thousands of years before being consumed by the sun. As it happened, the French astronomer U. J. J. LeVerrier (best known in connection with the discovery of Neptune) was at that time analyzing the discrepancies between computed and observed changes in the perihelion of Mercury, and exploring the possibility that they could be caused by small bodies moving between Mercury and the sun. It turned out that the amount of material needed, on Kelvin's theory, to maintain the heat of the sun would produce a much greater effect on Mercury's orbit than was in fact observed, and Kelvin was therefore forced to abandon the idea that the sun is maintained at a constant temperature by meteoric fuel. It is ironic that the same calculation of LeVerrier that led Kelvin to assume that the sun is cooling off revealed a fundamental flaw in Newtonian celestial mechanics, though the true significance of the advance of the perihelion of Mercury was not understood until Einstein proposed his general theory of relativity more than half a century later.

Kelvin mounted his attack on geology in earnest in the 1860s, first giving an estimate of the rate of cooling of the sun: 100°C during a period between 700 and 700,000 years (the uncertainty being due to lack of knowledge of the actual physical state and chemical constitution of the sun). But Darwin in his *Origin of Species* (1859), had conjectured that certain geological processes such as the gradual removal of solid material from chalk cliffs by water might have been going on for as long as 300,000,000 years. Kelvin thought, on the contrary, that it is "most probable that the sun has not illuminated the earth for 100,000,000 years, and almost certain that he has not done so for 500,000,000 years. As for the future, we may say, with equal certainty, that inhabitants of the earth cannot continue to enjoy the light and heat essential to their life, for many million years longer, unless sources now unknown to us are prepared in the great storehouse of creation" (Kelvin 1862: 393). Again note the hedge, which makes it possible for a modern admirer of Kelvin to point out that he never claimed more than he could prove, and that he was well aware that an argument based on the science of one generation may prove to be worthless when viewed with hindsight.

At the same time Kelvin published an article, "On the secular cooling of the earth," in which he used his results from Fourier's heat equation; he asserted (1862b: 610) that "essential principles of Thermodynamics have been overlooked by those geologists who uncompromisingly oppose all paroxysmal hypotheses" and maintain that geological actions have not been more violent in the past than they are now. Kelvin thought that the temperature and rate of dissipation of heat in the solar system in general, and on the earth in particular, must have been considerably greater in past times than they are now; thus geological speculations assuming greater extremes of heat, more violent storms and floods, etc., in remote antiquity are more probable than those of the extreme uniformitarian school. By using data on the rate of increase of temperature with depth in the earth's crust, and extrapolating backwards in time using the heat-conduction equations, he estimated that a period of 100 million to 200 million years might have been required for the earth, assumed initially to be at a uniform temperature of 7000° to 10,000°F, to reach its present state. (These two temperatures are his minimum and maximum estimates of the melting point of rocks.)

It is ironic that Fourier in 1820 had given a similar formula, and suggested roughly the same data, from which a similar estimate for the age of the earth could easily have been deduced; but apparently Fourier thought that 200 million years was such an incredibly large number that it was not even worth writing down. At that time, in fact, 200 million years was so much longer than the time scales being discussed by geologists that Fourier's results could be used to argue that the cooling of the earth was too *slow* to have any significant effect on surface temperatures during geological history; this argument in turn permitted the acceptance of Louis Agassiz' "ice age" theory which assumed that the surface has become *warmer* in the recent past! By the 1860s, Kelvin could use the same estimate of 200 million years to argue that geological time scales must be *shortened*.

In 1865 Kelvin published a short article with the provocative title "The 'Doctrine of Uniformity' in Geology briefly refuted." He argued that the rate of loss of heat from the surface of the earth at the present is so great that it could not possibly have been effective for as long as 20 million years in the past, since that amount of heat would have been enough to heat an amount of rock equal to 100 times the mass of the earth by 100°C.

Apparently Kelvin was infected by the common fallacy that an established scientific theory can be immediately overthrown by citing a single devastating argument against it; in any case, there was no immediate reaction from the geologists to this "refutation" of their theory, despite Kelvin's assertion that his conclusion "suffices to sweep away the whole system of geological and biological speculation demanding an 'inconceivably' great vista of past time, or even a few thousand million years, for the history of life on the earth" (1897: 344). It was not until 1868, when he addressed the Glasgow Geological Society, that the geologists paid any attention to his criticisms, although Darwin was already worried by the implications of the dissipation principle for biological evolution (as is shown by some of his letters). Kelvin began his lecture "On geological time" with the sentence: "A great reform in geological speculation seems now to have become necessary." He suggested that theorems on the stability of planetary motions, established by mathematical astronomers at the end of the eighteenth century, had been "taken up somewhat rashly, and supposed to imply more than they really did with reference to the permanence of the solar system." His main example of geological thinking was a passage from John Playfair's *Illustrations of the Huttonian Theory of the Earth,* in which Playfair asserted that the earth's history is a series of "vicissitudes of decay and renovation" of which we see neither beginning nor end; "the Author of nature has not given laws to the universe, which, like the institutions of men, carry in themselves the elements of their own destruction. He has not permitted in his works any symptoms of infancy or of old age, or any sign by which we may estimate either their future or their past duration" (Playfair 1802: 119).

To demolish this viewpoint, Kelvin first pointed out that the theorems of the French astronomers, on which Playfair's statement is apparently based, admittedly ignore the effects of frictional resistance (as well as being approximations based on first-order perturbation theory). He then reviewed the evidence for the effect of the tides on the rotation of the earth and moon, which he said indicated that the rotation has been gradually slowing down. Next, Playfair seemed to be assuming that the sun can go on forever providing us with heat at a constant rate, which is a violation of the laws of physics. Finally, Kelvin reiterated his arguments based on underground heat.

The first direct reply to Kelvin's attack on geology was given by T. H. Huxley in his Presidential Address to the Geological Society of London in 1869. Huxley quoted Kelvin's opening sentence (see above) and his further statement that "it is quite certain that a great mistake has been made—that British popular geology at the present time is in direct opposition to the principles of Natural Philosophy" (Kelvin 1894, 2: 44). Huxley appointed himself attorney general for the geologists, with the hope that although Kelvin's criticisms "involve the consideration of matters quite foreign to the pursuits with which I am ordinarily occupied" he was in this respect no worse off than most lawyers who "nevertheless strive to gain their causes, mainly by force of mother-wit and common sense, aided by some training in other intellectual exercises" (Huxley 1896: 306).

This was clearly the wrong approach to take, for it indicated to Kelvin and other physicists that Huxley was not only unable to refute their arguments based on physics and mathematics, but thought that he could win his "case" without giving any serious consideration at all to those arguments. At the same time Huxley tried to evade the issue by declining to defend uniformitarianism, and claiming instead that modern geology did not have to rely on long periods of past time to explain the present face of the earth:

> I do not suppose that, at the present day any geologist would be found to maintain absolute Uniformitarianism, to deny that the rapidity of the rotation of the earth *may* be diminishing, that the sun *may* be waxing dim, or that the earth itself *may* be cooling. Most of us, I suspect, are Gallios, "who care for none of these things," being of opinion that, true or fictitious, they have made no practical difference to the earth, during the period of which a record is preserved in stratified deposits (Huxley 1896: 326–27).

In other words, the researches on the past history of the earth and sun, on which Kelvin and other physicists had expended many years of labor, were of no interest to geologists. Huxley then tried to show that Kelvin's estimates are inconsistent and therefore could be discounted by geologists; in this he was apparently confused by the concept of upper and lower limits of an estimate based on the limits of uncertainty of the original data, and by the fact that different hypotheses result in different estimates. He ignored the fact that Kelvin had in one paper chosen the hypothesis most favorable to uniformitarianism in order to show that even with such an assumption the upper age limit is still too small, while elsewhere he had made a more realistic assumption in order to obtain a reliable estimate.

Just a month after Kelvin's address to the Geological Society of Glasgow, Archibald Geikie presented to the same society a paper "On modern denudation" in which he reviewed the evidence for the rate of removal of materials from the surface of land by rivers and other causes, and the bearing of this evidence on geological time. His principal conclusion is that the rate of denudation, when estimated by presently available evidence, is *not* "inconceivably slow" as Darwin and others had supposed. On the contrary, denudation continued at the present rate would suffice to wash away all the solid land on this planet in a few million years. Referring to Kelvin's address, he says: "We have been drawing recklessly upon a bank in which it appears there are no further funds at our disposal. It is well, therefore, to find that our demands are really unnecessary" (Geikie 1871: 189).

Geikie's moderate tone did not pacify Kelvin. Having finally succeeded in flushing out a defender of modern geology (Huxley), he eagerly pounced on him to tear him apart. His rejoinder to Huxley was delivered to the Geological Society of Glasgow in April, 1869, under the title "Of Geological Dynamics." He first attacked Huxley's legalistic style, saying that "the very root of the evil to which I object is that so many geologists are contented to regard the general principles of natural philosophy, and their application to terrestrial physics, as matters quite foreign to their ordinary pursuits" and that Huxley's sophistry would never satisfy "the high court of educated scientific opinion (Kelvin 1894, 2: 74–75). He gave a series of quotations from Darwin and from current geology textbooks to support his contention that geological speculation has in fact demanded very long periods of time. Geikie's paper mentioned above is applauded as a "secession from the prevailing orthodoxy" (ibid.p.87); he implied that this work, as well as Huxley's own revised and much smaller estimates of time periods, are the direct result of his (Kelvin's) proof that the longer estimates are untenable.

While Geikie's work was probably largely completed before Kelvin's 1868 address, it is certainly true that Kelvin's criticisms forced the geologists to be much more careful in estimating the time periods required for various processes, and also stimulated the collection of experimental data which could make such estimates more reliable (Geikie 1899: 722). But at the same time the dispute created a great deal of bitter feeling between physicists and naturalists, the effects of which can still be noticed today (see next chapter). The physicists' viewpoint was expressed by an article in the *North British Review* in 1869, entitled "Geological Time." The author quoted an oral remark of "a great living geologist" to the effect that he would not accept any result based on physics concerning geology if it were inconsistent with the results obtained by geologists using their own methods. The physicist author saw this attitude as similar to "those most objectionable theories and practices of the Trades-Unionists which have recently been held up to public execration"; the geologists had tried to rope off their own domain and would not listen to what anyone else has to say about it. "This sort of thing won't do in Science," said the physicist, "and the sooner scientific men of every species recognize the fact the better. . . . Mathematics are indispensable to the complete development of every real science" (1869: 407–9). Unfortunately British scientists have not yet realized this, although the situation is quite different in other countries, he noted.

Naturally one would like to know the identity of this anonymous reviewer who supported Kelvin in his attack on the uniformitarian geologists, as well as berating natural scientists in general for their mathematical ignorance. The article made a strong impression on Darwin, who wrote to Hooker that it is "admirably done" though containing "some good specimens of mathematical arrogance"; he observes that "geologists have all been misled by Playfair, who was misled by two of the greatest mathematicians," which perhaps only goes to show "how cautious geologists ought to be in trusting mathematicians." Later his son, George Darwin, convinced him that the article was probably written by P. G. Tait, the Scottish mathematical physicist who was coauthor with Kelvin of the well-known *Treatise on Natural Philosophy* ([Darwin, Charles] 1903, 1: 313–14).

Although the main brunt of Kelvin's attack was borne by geology, it is clear that Kelvin and other physicists were somewhat hostile to Darwin's theory, and an important by-product of the calculation of the rate of dissipation of heat by the earth and sun was the conclusion that the physical conditions on the earth's surface could not have been favorable for the development of life over a sufficiently long period to permit evolution by natural selection alone. Kelvin (1869) stated that

> The limitation of geological periods, imposed by physical science, cannot, of course, disprove the hypothesis of transmutation of species; but it does seem sufficient to disprove the doctrine that transmutation has taken place through "descent with modification by natural selection" (Kelvin 1894, 2: 89–90).

Kelvin's opposition to Darwin's theory has been somewhat exaggerated by historians. His writings show that while he did not object to the principle of evolution itself, he did dislike the aspect of randomness and lack of conscious direction implied by the hypothesis of natural selection. In his Presidential Address to the British Association in 1871, he agreed with the astronomer Sir John Herschel that Darwin's proposed mechanism for evolution was

> too like the Laputan method of making books, and that it did not sufficiently take into account a continually guiding and controlling intelligence. . . . The argument of design has been greatly too much lost sight of in recent zoological speculation[;] . . . overpoweringly strong proofs of intelligent and benevolent design lie all round us . . . teaching us that all living things depend on one ever-acting Creator and Ruler (Kelvin 1894, 2: 204–5).

As mentioned above, Darwin was well aware of the fact that Kelvin's estimates of the age of the earth, if correct, would weaken his own theory of evolution; this is shown by his letters to Croll, Hooker, and Wallace in 1869. In 1871, he wrote to Wallace: "I should rely much on pre-Silurian times; but then comes Sir W. Thomson like an odious spectre" (Marchant 1916: 220). By this time his estimate of 300 million years for the denudation of the Weald, which had attracted Thomson's derision, had been removed from later editions of the *Origin of Species*. Loren Eiseley states that the physicists "forced Darwin, before his death, into an awkward retreat which mars in some degree the final edition of the *Origin*" (Eiseley 1958: 245).

Although Darwin retreated from the battle, the debate between physicists and geologists continued throughout the rest of the century. Huxley, lecturing on evolution in New York in 1876, considered the physical argument entirely irrelevant to biology. It is up to the geologists and physicists, he declared, to decide on the age of the earth; once they have agreed among themselves, biologists will accept the decision. Biologists are basically interested only in whether it is a fact that evolution did take place. "We take our time from the geologists and physicists; and it is monstrous that, having taken our time from the physical philosopher's clock, the physical philosopher should turn round upon us, and say we are too fast or too slow" (Huxley 1894: 134). This seems to me a reasonable argument, in view of the fact that evolutionary theory did not at that time *require* any particular time scale, in the absence of direct *biological* evidence of the *rate* of evolutionary change. Huxley's point was ignored by those whose pulses raced at the prospect of a decisive battle between physics and evolution.

The geologists did revise their time-scale so as to make it fall within Kelvin's original limit of 100 million years, but at the same time physicists such as Tait and Clarence King reduced that limit to somewhere between 10 and 20 million years. While the physicists still maintained that there had been insufficient time for uniform geological processes and evolution by natural selection, the geologists, while willing to abandon the extreme uniformitarian viewpoint, became more and more irritated by the refusal of physicists to give any weight to geological evidence. As Geikie said in his British Association address in 1899, "It is difficult satisfactorily to carry on a discussion in which your opponent entirely ignores your arguments, while you have given the fullest attention to his" (Geikie 1899: 724).

Despite Huxley's insistence that geophysical studies could not alter the fact of biological evolution but only its time scale, a group of American naturalists argued that the theory of evolution would be improved by admitting a possible effect of environment. Clarence King, soon to become the first Director of the U. S. Geological Survey, discussed the possible relations of catastrophism and evolution in a lecture at Yale in 1877. Granting that Darwin and Huxley were not fanatic uniformitarians, he attacked "lesser men who bang all the doors, shut out all doubts, and flaunt their little sign, 'Omniscience on draught here.' It must be said, however, that biology, as a whole, denies catastrophism in order to save evolution" (1877:463-64). But, said King, when catastrophes change the environment, only the more "plastic" forms of life are able to survive. This responsiveness of plastic species to environmental effects would speed up the process of evolution. "Moments of great catastrophe, thus translated into the language of life, become moments of creation, when out of plastic organisms something newer and nobler is called into being" (1877: 470).

Others—E. D. Cope, Alpheus Hyatt, and Alpheus Packard in America and Ernst Haeckel in Germany—were expressing similar or more complicated ideas which are now known by the general term "neo-Lamarckism." It is not necessary to follow the history of this subject except to note that along with the idea of accelerated evolution, which could help make biology conform to the physical time scale, went also the converse idea of retarded evolution, which was already being used in a slightly different sense in the theory of degeneration (see Chapter VII).

The outcome of the dispute about the age of the earth, which is now well known, can be summarized briefly. At the beginning of the twentieth century the situation changed radically for a reason that neither geologists nor physicists could have anticipated—the discovery of radioactivity. Himstedt pointed out in 1904 that if there are deposits of radium (recently isolated by the Curies) in the earth, then the heat generated by radioactivity must be taken into account in studies of the thermal history of the earth. Soon afterward, Liebenow followed up this suggestion by estimating that the presence of 1/5000 of a milligram of radium per cubic meter, distributed uniformly throughout the earth's volume, would be sufficient to compensate for the observed loss of heat by conduction through the crust. A similar suggestion was made about the same time by Rutherford and Strutt. Thus the possibility of continual generation of heat over long periods of time invalidated Kelvin's assumption that the earth is simply cooling down from an initial molten state.

It was soon recognized that the relative proportions of lead, helium, radium, and uranium in rocks could be used to estimate the ages of those rocks. Strutt, in 1905, obtained the estimate 2.4×10^9 years by measuring helium and radium; Boltwood, in 1907, found 2.2×10^9 years for one rock sample by measuring the lead-uranium ratio. The inspiration for both studies seems to have come from Rutherford. After some initial skepticism the validity of this method for estimating ages was generally recognized, along with the thousandfold increase in the time scale over that which had previously been accepted. The geologists and biologists were of course delighted by this turn of events, while the physicists were too busy exploring the exciting new domain of radiation and nuclear transformations to pay much attention to reproaches from the naturalists.

Lord Kelvin lived long enough to see this reversal of opinion; according to J. J. Thomson (1937: 420), "he told me that before the discovery of radium had made some of his assumptions untenable, he regarded his work on the Age of the Earth as the most important of all." We have Rutherford's own account of a lecture which he gave at the Royal Institution in 1904, in which he discussed the influence of radioactive heating on estimates of the earth's age:

> I came into the room, which was half dark, and presently spotted Lord Kelvin in the audience and realized that I was in for trouble at the last part of my speech dealing with the age of the earth, where my views conflicted with him. To my relief, Kelvin fell asleep, but as I came to the important point, I saw the old bird sit up, open an eye and cock a baleful glance at me! Then a sudden inspiration came, and I said Lord Kelvin had limited the age of the earth, provided no new source was discovered. That prophetic utterance refers to what we are now considering tonight, radium! Behold! The old boy beamed upon me   (Eve 1939: 107).

The Kelvin-Huxley dispute has been largely forgotten, and twentieth-century scientists may well ask why we have bothered to resurrect it here. True enough, modern geologists can afford to ignore this episode; after all, Kelvin was wrong. Likewise, biologists nowadays have little interest in the theory of degeneration, which we discuss in Chapter VII; it was a dead end, and was bypassed in the development of modern evolutionary theory. But the historian can learn much from the failures as well as from the successes of science. The relations between different sciences, and between science and culture, are more strongly affected by the transference of ideas, and their successful or unsuccessful application to new problems, than by the mere occurrence of great discoveries. The historical role of individual scientists is also revealed more clearly by examining their adventures in fields outside their own specialty; being unfamiliar with the jargon, they are forced to express themselves in plain language, and they often expose their philosophical preconceptions. Thus Kelvin, who is frequently cited as the foremost advocate of mechanical models in nineteenth-century physics, turns out to be an opponent of mechanism in geology and biology. His particular criticism was typical of the neoromantic attack on realism, even in its use of mathematics and experimental data to refute a theoretical generalization.

Chapter IV

Planetary Science: From Underground to Underdog

> ... earth-lore is not a discrete science at all, but is that way of looking at the operations of energy in the physical, chemical and organic series which introduces the elements of space and time into the considerations and which furthermore endeavors to trace the combination of the various trends of action in the stages of the developments of the earth. It is in these peculiarities of geology that we find the basis of its value in education and in the general culture of society.
> —(Shaler 1896: 319)

It is now time to step back from our detailed examination of nineteenth-century controversies and give some consideration to another kind of change which has taken place in the structure of science during the past century, and to see how such changes affect fashions in the historiography of science.

When a historian looks at the science of earlier centuries, he generally starts with some preconceptions about what kinds of theories and experiments are important enough to merit detailed examination. In recent years historians of science have attempted to avoid the "whig" or "present-minded" attitude—the idea that one should concentrate on those events that somehow led to the establishment of today's accepted laws. Instead we have seen an increasing emphasis on understanding the scientific activity of each historical period on its own terms, with due regard to the context of theories and problem situations.

Nevertheless there still remains a major source of distortion in our historical interpretations, a more general kind of present-mindedness, namely, the influence of the level of importance or prestige that is *now* ascribed to one area of science in relation to others which may be different from the situation in the period we are studying. I want to present here a particular example of this: the so-called *planetary* sciences as contrasted to the so-called *pure* sciences.

By planetary sciences I mean the study of the earth and its atmosphere, and the rest of the solar system (including the sun). By pure sciences I mean the study of those properties of matter and energy that are presumably applicable everywhere in the universe.

I propose the following assertions:

1. Until quite recently, planetary science has been assigned an inferior role in contemporary disciplines such as physics, so that regardless of how much money may be available for such research it has not carried the prestige (among scientists themselves) associated with more "fundamental" (pure) areas such as elementary particles, cosmology, or even solid-state physics and molecular biology. As a result, many competent scientists who might well have made important discoveries in planetary science did not even consider entering this field.
2. Before the nineteenth century, the situation was quite different: planetary science was regarded as an integral part of science, with as much significance and social status as any other part; as a result, many of the greatest scientists devoted considerable effort to planetary problems.
3. Problems in planetary science frequently provided a stimulus for discoveries in what we now define as pure science; in this chapter I will be concerned primarily with examples in the nineteenth century.
4. During the nineteenth century the distinction between pure and planetary science (as defined above) began to develop; this was in part due to the inevitable specialization and profes-

sionalization of all the sciences that occurred during this period, and is reflected in the initiation of new societies and journals for the planetary sciences. But that is not sufficient to account for the development of scientists' attitudes concerning the relative value or fundamental character of the different disciplines; here one must pay attention to the role of particular scientists and events.

5. The twentieth-century historian of science has too often accepted the physicists' assumption that planetary science is inferior to pure science, and has allowed this assumption to prejudice his interpretation of the science of earlier centuries. This bias has led historians to ask pseudo-questions such as, "Why were Americans indifferent to basic science during the nineteenth century?" or "Why did a brilliant scientist who was capable of making great discoveries in pure science waste his time on problems in planetary science?" The wording of such questions implies that planetary science (in which, for example, the Americans excelled in the nineteenth century) is not as worthwhile as pure science, or at least has nothing to contribute to it, and therefore is an occupation suited only to second-raters. But in fact many first-rate scientists—first-rate even if judged only on their contributions to what we now call pure science—devoted substantial effort to research in planetary science. The historian, if he hopes to get an accurate picture of the history of science before the twentieth century, must make some effort to understand why men like Dalton, Gauss, Kelvin, and Helmholtz thought planetary problems were important. An added benefit of such effort might be to open up some new areas of historical research for a profession that seems to me to have concentrated too much on the "mainstream" of pure science while neglecting some of the questions that earlier scientists considered important.

In this chapter I will deal primarily with planetary physics, so the evidence for my first assertion comes from an examination of attitudes expressed by modern physicists. I think it is relevant to point out that I used to share these attitudes myself. In my own education in physics in the late 1950s, I was never made aware that planetary physics was part of physics; I absorbed all the prejudices of the physicists against people who worked in that area, if I ever thought about them at all. Physicists whom I knew as teachers and colleagues would just naturally advise their best students to go into fields like elementary particles, solid state, or cosmology; only someone who wasn't good enough to do "real" physics would think of going into planetary physics.

I don't recall noticing any marked change in these attitudes during the early 1960s, though I now realize that planetary physics was enjoying a sensational revival at that time, stimulated by the theories of continental drift, observations of sea-floor spreading, and upper-atmosphere experiments made possible by the space program. It was only around 1970, when cutbacks in government funding caused widespread panic and unemployment among scientists and engineers, that physicists began to wonder whether they should look into planetary science, on the assumption that money might be available for "environmental" research.

In September 1970, Freeman Dyson, a well-known mathematical physicist at the Princeton Institute for Advanced Study, wrote in *Physics Today* about new areas to which physicists might turn. He mentioned the study of "activated sludge" as an example of a problem to be solved in the "environmental field" but warned his reader that "he should not expect that what he does in the environmental field will be mainly physics. If he is any good, he will use his physics only as a cultural background in thinking about problems that are primarily chemical, biological or political in nature. Accordingly I think it would be a mistake for a physics department of a university to become heavily involved in environmental work" (1970: 26). This was as close as Dyson came to saying anything about planetary science as a subject that might be of interest to physicists.

As it happened, the same issue of *Physics Today* carried a book review by Fred Wilson, who recalled:

> An oral examination question attributed to Enrico Fermi is "How much energy is dissipated in a bolt of lightning?" The reaction of most examinees would be stark panic because most physics students learn little or nothing about the physics of

their environment. This is particularly sad because many advances in physics were intimately connected with the development of meteorological instruments (Wilson 1970: 53).

The follow February, physicist H. R. Crane wrote on "Opportunities in Geophysics" in the same journal; he suggested that the job crisis might force physicists to look to the earth sciences for employment but admitted that it would be distasteful for his readers to push themselves into an "applied science." Nevertheless,

> it will be up to physicists to recultivate geophysics and other areas of opportunities after having neglected them for a generation. Recultivation may require changes in the curriculum and research experience, perhaps reaching down into the physics-major program (Crane 1971: 26).

The clear implication was that, had it not been for the pressures of the marketplace, no respectable physicist would dream of getting involved in this field.

More recently, in response to some of my own preliminary explorations of the connections between pure and planetary sciences, I have elicited some frank opinions which the authors would probably prefer not to have publicly attributed to them:

> [Scientist A:] "Earth and Planetary Science" today is the "social science" among the real sciences. It is having a great go with the $$$, like the "interdisciplinary X" of a decade ago, and "aerospace" two decades ago. It is wonderful for deans and all those who didn't make the grade in some science or other.
> [Scientist B:] The real division is between "analytical scientists" and "collectors of data." Then (19th century) as now the former tended to look down on the latter, for they alone were in full control of their medium, knowing fairly well not only their tools but also the exact aims which they were pursuing. The poor data collectors, on the other hand—classifying their fauna and flora, cloud and wind patterns, geological strata and whatnot—hardly ever had a clear concept of what they

were after, and therefore rarely managed to extract ideas from their findings. . . .

Scientist B's dichotomy between analytical scientists and data collectors is quite pertinent to my theme here; I would argue that in principle there is room for both types of worker in any discipline, but that, since the nineteenth century, analytical scientists have largely deserted planetary science. Why have they done so?

The other aspect brought out by Scientist A is the feeling that planetary science boondoggles in the past decade or two have wasted enormous amounts of money that could better have been spent on pure science. Projects like manned exploration of the moon have been criticized as publicity stunts in which little scientific knowledge was gained that could not have been gleaned much more cheaply (and with less risk to human life) by unmanned probes.

When the "Physics Survey Committee" of the National Academy of Sciences published a report in 1966 for the purpose of extracting more money from the government they were very careful to exclude planetary physics from their definition of physics, although they were willing to include related areas of astronomy, chemistry, and engineering. I interpret this as a message that the planetary sciences have already gotten more money than they deserve from the government—if the government gives them any more it should not be under the illusion that it is supporting pure science by doing so!

A more recent (1972) report from the same committee recognized the existence of planetary science but tried to blame Aristotle for its low status:

> Aristotle divided the physical sciences into physics and meteorology, the latter embracing all beneath the orbit of the moon and corresponding roughly to what we now call physics of the earth. He understood that physics would embrace the more orderly branches of nature but that meteorology would have to deal with complicated phenomena in which descriptive methods, as opposed to the more abstract processes of induction and deduction, would play an essential role.

In the last two centuries, conventional physics and studies of the sea, atmosphere, and earth have drawn apart. New experimental methods led physics rapidly toward goals concerned with the fundamental nature of matter and radiation, while industrial needs coupled to new technological advances, such as the telegraph and deep-drilling rigs, led studies of the earth to increasingly technical and descriptive methods, with little emphasis on fundamental understanding. These trends are now beginning to reverse, and the proper role of physics in earth and planetary investigation is becoming clearer (Physics Survey Committee 1972: 261).

A more positive view of recent history was presented at about the same time by Knopoff:

A century ago geology—under the impact of the theory of evolution, the statement of the principles of stratigraphy and widespread exploration—was perhaps the most exciting area of science. With passing years, geological activities were overshadowed by dramatic discoveries in chemistry, physics, biology, and astronomy. The recent new discoveries relating to the solid earth: i) put earth sciences once again in the forefront; ii) unified previously diverse fields of geology and geophysics; and iii) greatly enhanced the morale of geologists and geophysicists. The new excitement in the earth sciences is attracting many talented young people into this complex but highly important area of science (Knopoff 1972: 4–5).

Since then, planetary physics has continued to move into the center of scientific attention. But there are still many physicists who believe, with the Physics Survey Committee, that the planetary sciences have historically been restricted to "descriptive methods, with little emphasis on fundamental understanding." That misconception has been adopted by historians of modern physics, who have concentrated on studying the development of atomic models, relativity, and electromagnetic field theory. Historians would probably accept my second assertion, that "before the nineteenth century, . . . planetary science was regarded as an integral part of science, with as much significance and social status as any other part; as a result, many of the greatest scientists devoted considerable effort to planetary problems." It requires only the most superficial glance at the works of giants such as Galileo, Newton, Euler, Lagrange and Laplace to see that they considered problems such as the motions of the planets, the figure of the earth, and the tides, to be of fundamental importance. But when historians of science move into the nineteenth century, they seem to consider geology a subject that is interesting only as part of the background for Darwin's work; other areas of planetary science are scarcely noticed at all. Of course people have written widely on the history of geology, meteorology, and planetary and solar astronomy as independent sciences, but only a handful of historians (Nathan Reingold, Harold Burstyn, Joe Burchfield, Elizabeth Garber, and a few others) have given serious consideration to the historic connections between geophysics and "pure" physics (see the conference report, Gillmor 1975).

The study of the thermal history of the earth, which led to the introduction of the principle of irreversibility into physics by Fourier and was one of the two sources for Kelvin's statement of the general principle of dissipation of energy, is an excellent example of the progress in pure physics which has been stimulated by problems in planetary physics. Fourier himself stated quite clearly in 1827 that such problems had motivated his own work: "The question of terrestrial temperatures has always appeared to me as one of the greatest objects of cosmological studies, and I had it chiefly in view when I established the mathematical theory of heat conduction" (Fourier 1890: 114).

Studies of the nature of the atmosphere also interacted with the growth of physics and chemistry, beginning in the seventeenth and eighteenth centuries with the concepts of air pressure and temperature and the development of instruments to measure them. The discovery that air consists of more than one kind of gas led to major changes in chemical theory at the end of the eighteenth century. John Dalton, in the early research which led to his formulation of the chemical atomic theory (1803), tried to explain why oxygen and nitrogen remain mixed in constant proportions in the mosphere despite their difference in density. He guessed that it had something to do with the differences in sizes or weights of the atoms of different elements, and was then led to investigate chemical methods for determining such differences. He was impressed by the results of a balloon flight, showing that the air high above Paris has the same composition as that near the ground. The outcome of Dalton's work was the first table of atomic weights (Manley 1968; Thackray 1972).

J. L. Gay-Lussac, one of the French scientists who undertook these balloon ascents to study the atmosphere, is known for two major discoveries: the law of combining volumes, and the law of thermal expansion of gases. The first of these emerged from the attempt to determine the chemical composition of the atmosphere (Crosland 1961); the second was in part the result of a request from Laplace for information about the physical properties of the atmosphere needed to make refraction corrections to astronomical observations.

Rudolf Clausius worked on the problem of light-scattering in the atmosphere in the 1840s; in this connection he developed some of the statistical methods which he later used in the kinetic theory of gases (Schneider 1974).

Another atmospheric problem was the variation of temperature with respect to height above the earth's surface. Throughout most of the nineteenth century, the available data suggested that there is a uniform linear decrease of temperature with height. Such a simple empirical law cried out for a simple theoretical explanation, and the pioneers of the kinetic theory of gases, John Herapath and J. J. Waterston, tried to give one. This might have turned out to be one of the successful applications of the kinetic-molecular theory of matter, establishing another link between pure science and planetary science. Instead, it was found that both the observational data and the theoretical arguments were misleading. Maxwell and Boltzmann showed that the equilibrium state of a vertical column of gas is characterized by uniform temperature throughout; gravity by itself in the absence of any heat source at the top or bottom of the column cannot produce a temperature gradient. Some years later, around 1900, balloon experiments at higher altitudes showed that the temperature of the atmosphere does not continue to decrease with height but actually increases in some regions. The eventual convergence of theory and observation came too late to have any significant effect on the main lines of development of either pure or planetary science.

The atmospheric temperature problem did produce one contribution to the debate on the Second Law of Thermodynamics. Josef Loschmidt, Boltzmann's colleague at Vienna, objected to the theoretical conclusion that thermal equilibrium in a vertical column of gas requires uniform temperature throughout, and thought he could refute not only this conclusion but the more general one that thermodynamics implies a "heat death," with uniform temperature throughout the universe. One of his arguments was the "reversibility paradox"—all molecular velocities could be reversed without violating Newton's laws, and the system must then return to its initial state, contrary to the principle of irreversibility. Boltzmann's reply led directly to his statistical theory of entropy (see next chapter).

Going beyond geophysics to the solar system, one finds two other problems that had close connections with nineteenth-century heat theory. The first is the structure of Saturn's rings. Maxwell showed in 1856 that no reasonable form of *solid* ring could be mechanically stable, whereas a satisfactory theory could be based on the hypothesis that the rings are composed of swarms of small particles. From this research he turned to the theory of gas viscosity (which also involved interactions of streams of particles) in 1865, and eventually developed the kinetic theory of transport processes.

The other problem related to heat theory was that of estimating the temperature of the sun. Yielding a plausible estimate was one of the criteria used to evaluate laws of radiative heat transfer in the nineteenth century (Brush 1973). In 1817, P. L. Dulong and A. T. Petit proposed an exponential law for the rate of heat loss by a hot body, as a replacement for Newton's law of cooling. Their law was supported by impressive experimental evidence, and was generally accepted for the next fifty years. However, it was eventually rejected because estimates of the temperature of the sun based on it led to an absurdly low value (about 1400°). Because a number of physicists took an interest in determining the temperature of the sun (and in the related problem of the transmission and absorption of radiant heat by the earth's atmosphere), new experiments on the radiation law were undertaken. The outcome was Stefan's suggestion that the data could best be represented by assuming that the rate of emission of energy is proportional to the 4th power of the absolute temperature. With the aid of further experiments and Boltzmann's theoretical derivation of this law, the T^4 formula was established in the 1880s, and played a role in the subsequent search for the law governing the frequency-distribution of black-body radiation (and thus in the genesis of Planck's quantum theory).

It is also of interest to inquire whether the scientist who makes such a discovery or theoretical development in pure science then turns his back on planetary science, choosing to concentrate on following up the consequences of the new work, or whether he regards the planetary problem as sufficiently important to be worth further study. Kelvin certainly chose the latter course, publishing many later papers on geophysical topics. It is not so well known that Maxwell, after developing his kinetic theory of gases, did return to the problem of Saturn's rings, and (in manuscripts soon to be published) applied his new kinetic methods to the problem.

With these and other examples of fruitful interactions between planetary and pure science in mind, one may well ask why they should have become separated, and why the idea should have arisen that the planetary sciences are not as important as the pure sciences.

The process of separation can be studied on a quantitative basis by examining the distribution of research papers in specialized as opposed to general-science or general-physical-science journals. It can also be analyzed in the bibliographies of selected major scientists, by asking what role research in the planetary sciences played in their entire output. But there is obviously a more subjective aspect involved in the development of scientists' attitudes concerning the relative value or fundamental character of the different disciplines. I think this historical trend is best represented by the writings of three nineteenth-century thinkers: August Comte, Lord Kelvin, and Max Planck.

Comte, in the first volume of his *Course of Positive Philosophy* (1830) made a sharp distinction between abstract sciences which attempt to discover laws, and concrete descriptive sciences ("natural sciences") which apply those laws. The former are "fundamental" (e.g., physiology and chemistry); the latter include zoology, botany, and mineralogy. Each concrete science presupposes the study of the abstract ones; thus the special study of the earth requires knowledge of physics and chemistry, and also of astronomy and physiology.

By 1887 Comte's system of classification (even if not entirely original) had been so widely adopted that the anthropologist Franz Boas, who himself had started out his career as a physicist, had to defend the descriptive natural sciences against the tendency to exalt the fundamental sciences:

> The fact that the rapid disclosure of the most remote parts of the globe coincided with the no less rapid development of physical sciences has had a deep influence upon the development of geography; for while the circle of phenomena became wider every day, the idea became prevalent that a single phenomenon is not of great avail, but that it is the aim of science to deduce laws from phenomena; and the wider their scope, the more valuable they are considered. The descriptive sciences were deemed inferior in value to researches which had hitherto been outside their range. Instead of systematic botany and zoology, biology became the favorite study; theoretical philosophy was supplanted by experimental psychology; and, by the same process, geography disintegrated into geology, meteorology, etc.
>
> ... The physical conception is nowhere else expressed as clearly as in Comte's system of sciences. ... The single phenomenon itself is insignificant; it is only valuable because it is an exemplification of a law, and serves to find new laws or to corroborate old ones. To this system of sciences Humboldt's "Cosmos" is opposed in principle. Cosmography, as we may call this science, considers every phenomenon as worthy of being studied for its own sake. ... While physical science arises from the logical and aesthetic demands of the human mind, cosmography has its source in the personal feeling of man towards the world, towards the phenomena sur-

rounding him. We may call this an "affective" impulse, . . . the physiognomy of the earth . . . cannot afford a satisfactory object of study to the physicist, as its unity is a merely subjective one (Boas 1887: 137–38, 140–41).

It was Kelvin who insisted that the geologists were wrong to suppose that the age of the earth was several hundred million years, since, no matter how much *geological* evidence they might find for that assumption, it was contradicted by the principles of physics (see previous chapter). What is surprising is not that Kelvin, as a physicist, thought geological evidence to be of little value against physical arguments based on heat conduction theory, but rather that the geologists themselves accepted that judgment. Leonard Wilson (1969) states that, despite Rutherford's refutation of Kelvin by radioactive evidence,

> The damage done to geology during the period from 1865 to 1903 was not to be so quickly remedied. A whole generation of geologists had grown up uncertain of their principles in theoretical geology. Lyell had been discredited, at least so far as the rate and intensity of geological causes were concerned, and geologists felt free to invoke catastrophic explanations of past events. . . . The influence of Lord Kelvin on geological thought over the past century remains profound and will not easily be erased. One of its consequences has been to obscure the historical significance of Charles Lyell and his concept of the uniformity of nature (Wilson 1969: 430).

One measure of Kelvin's success in undermining the self-confidence of geologists is the fact that even after radioactivity had discredited Kelvin's theory of the *age* of the earth, they continued to accept physical arguments for its *rigidity*. Kelvin, following his Cambridge teacher William Hopkins and with additional support from George Howard Darwin (son of Charles), attacked the earlier notion that the inside of the earth is mostly liquid. The position of "the physicists" at the end of the nineteenth century was that the earth is mostly if not entirely solid, and that it behaves like a globe more rigid than glass. While many geologists still believed that geological observations could be more easily explained with the assumption of a thin deformable crust over a liquid interior, they "reluctantly abandoned" this idea (Geikie 1903), especially after the publication of seismic wave observations which showed that "the earth throughout the greater part of its mass is capable of transmitting two types of elastic waves, and is therefore an elastic solid" (Knott 1899: 577).

When Alfred Wegener proposed his hypothesis of continental drift in 1912, he could make little headway against the prevalent belief in a rigid earth. R. D. Oldham described one of the major obstacles to acceptance of Wegener's theory in a discussion held in 1923:

> I can remember that when I started as a geologist the idea was not unknown that a good deal of geological evidence was, at any rate, not inconsistent with the notion that the continents have not always occupied the positions on the surface of the globe which they do now. But also I can remember very well that in those days it was unsafe for anyone to advocate an idea of that sort. The physicists, who before that had forced on us the notion of a fiery globe with a molten interior and thin crust on it, had gone round and insisted on a solid heated sphere, and they would allow us to appeal to nothing, as the cause for various structures and changes that we knew in geology, but the slow cooling and contraction of this solid globe, and any notion of the shifting of continents was incompatible with that theory. Those ideas held the ground so strongly that it was more than any man who valued his reputation for scientific sanity ought to venture on to advocate anything like this theory that Wegener has nowadays been able to put forward (Oldham 1923: 189).

Wegener himself, in 1927, mirrored the contemporary feeling that physical evidence is superior to geological evidence:

> I believe that the final resolution of the problem can only come from *geophysics*, since only that branch of science provides sufficiently precise methods. Were geophysics to come

to the conclusion that the drift theory is wrong the theory would have to be abandoned by the systematic earth science as well, in spite of all corroboration, and another explanation for the facts would have to be sought (Wegener 1966: vii).

While Kelvin wanted to bring physics into planetary science, Max Planck (1909) wanted to purge physics itself of "anthropomorphous" aspects that appeared to restrict its validity to one part of the universe. He argued that the laws of physics must have a universal character, involving constants that are independent of the properties of any special substance:

> Thus they can be used to determine units of length, time, mass, and temperature, and these units must necessarily retain their meaning for all time and for all extraterrestrial and superhuman *kulturs*. It is known that this does not, by any means, hold for our ordinary weight units. Though these are usually described as absolute units, it must be borne in mind that they bear special relation to our present terrestrial life. The centimetre is derived from the present circumference of our planet, the second from its time of rotation, the gramme from water as the principal constituent of the earth's surface, and temperature from the characteristic points of water (Planck 1960: 18).

Though Planck's philosophical standpoint was very different from that of Comte, he campaigned to establish the superiority of what Comte had called "fundamental" sciences over descriptive ones. Planck's quantum theory—later developed into quantum mechanics—and Einstein's relativity theory became paradigms of fundamental theories dominating all of science.

The decline of geology relative to physics was accelerated by the flowering of atomic physics in the first decades of the twentieth century. The fact that physicists had reached an incorrect conclusion about the age of the earth during the previous century was of little significance compared with the fact that they had now acquired very powerful tools which could be used to investigate this and many other problems. Lacking their own reliable methods the geologists either had to learn physics or waste their time in sterile controversy and data-collecting.

While I have not yet looked in detail at the profession of geology in the twentieth century and therefore cannot make any very definite statements about why it (and the planetary sciences in general) lost so much prestige within the the scientific community, some confirmation of the overall picture is given in the very interesting book recently published by Henry Menard (1971). According to Menard, geology was moribund during the period from about 1860 to about 1940 because it lacked the techniques needed to solve its important problems. "The pioneers of the nineteenth century identified big problems and, if they were solvable with the available tools, they solved them. The geologists who followed, presumably equally energetic and intelligent, were inevitably doomed to working on trivia until new tools were forged" (p. 144). Geologists in the twentieth century became accustomed to carrying on interminable controversies about problems they were unable to solve. Taking geology in this period as a paradigm of "dormant science," Menard notes that "The English deteriorates as concern with style grows. Jargon flourishes. An increasing fraction is literature of literature and bibliographies grow longer, and citations grow older" (p. 145). These are statements he has attempted to document at least in part by extensive statistical analysis. He also says that the publication time grew longer, from two or three months after receipt of a manuscript in 1890 to eighteen months in 1950 for the *Bulletin of the Geological Society of America*. The International Geological Congress demanded abstracts of papers two years in advance of meetings. The result of all this bureaucratic inertia was that when the revival did come in the earth sciences, the new practitioners refused to be called "geologists" but instead organized various hyphenated disciplines.

Chapter V

The Heat Death

> One hears it often said that in this world everything is a circuit. While in one place and at one time changes take place in one particular direction, in another place and at another time changes go on in the opposite direction; so that the same conditions constantly recur, and in the long run the state of the world remains unchanged. Consequently, it is said, the world may go on in the same way for ever. . . .
> The second fundamental theorem of the mechanical theory of heat contradicts this view most distinctly . . . in all the phenomena of nature the total entropy must be ever on the increase . . . *The entropy of the universe tends toward a maximum.*
> The more the universe approaches this limiting condition in which the entropy is a maximum, the more do the occasions of further changes diminish; and supposing this condition to be at last completely attained, no further change could evermore take place, and the universe would be in a state of unchanging death.
> —(Clausius 1868: 417–19)

To what extent did the physical principle of dissipation of energy actually influence nineteenth-century writers on philosophical or cultural subjects? Aside from Kelvin's own application of his principle to geological history, there was remarkably little impact on European thought until the end of the nineteenth century. It is only around 1900 that we find an increasing number of references to the second law of thermodynamics, and attempts to connect it with general historical tendencies. Whereas popular expositions of physics in the nineteenth century, such as Tyndall's *Heat as a Mode of Motion* (1863), made scarcely any reference to the implications of the second law, the more recent essays of Sir James Jeans (1929,1933) and Sir Arthur Eddington (1928) have made the "heat death" an integral part of the modern educated layman's knowledge of cosmology. Theologians as well as philosophers have had to take thermodynamics into account (Hiebert 1966, 1967).

The English philosopher Herbert Spencer was one of the first writers to take up the idea of dissipation of energy and incorporate it into a general system of philosophy. In his *First Principles* (1862) he made extensive use not only of evolution and the conservation of energy but also of equilibration and dissolution. Spencer gave several examples of equilibration in mechanical processes: "Every motion being motion under resistance, is continually suffering deductions; and these unceasing deductions finally result in the cessation of the motion." Discussing the equilibration of the molecular motion known as heat, he says:

> The tacit assumption hitherto current, that the Sun can continue to give off an undiminished amount of light and heat through all future time, is fast being abandoned. Involving as it does, under a disguise, the conception of power produced out of nothing, it is of the same order as the belief that misleads perpetual-motion schemers. The spreading recognition of the truth that force is persistent, and that consequently whatever force is manifested under one shape must previously have existed under another shape, is carrying with it a recognition of the truth that the force known to us in solar radiations, is the changed form of some other force of which the Sun is the seat; and that by the gradual dissipation of these radiations into space, this other force is being slowly exhausted. . . . Infinitely remote as may be the state when all the motions of masses shall be transformed into molecular motion, and all the molecular motion equilibrated; yet such a state of complete integration and complete equilibration, is that towards which the changes now going on throughout the Solar System inevitably tend. . . . If Evolution of every kind, is an increase in complexity of structure and function that is incidental to the universal process of equilibration, and if equilibration must end in complete rest; what is the fate towards which all things tend? . . . If Man and Society are similarly dependent on this supply of force that is gradually coming to an end; are we not manifestly progressing towards omnipresent death? (Spencer 1958: 487, 489, 507–8).

Spencer claimed to be able to prove that this general tendency toward dissipation is a direct consequence of the law of conservation of energy. His "deduction" consisted in asserting that physical phenomena can be interpreted only as the results of attractive and repulsive forces between atoms that fill all space; hence all motion is motion under resistance, and thus subject to equilibration. But this is not all that can be deduced from the conservation of energy by a clever philosopher:

> By this ultimate principle is provable the tendency of every organism, disordered by some unusual influence, to return to a balanced state. To it also may be traced the capacity, possessed in a slight degree by individuals, and in a greater degree by species, of becoming adapted to new circumstances. And not less does it afford a basis for the inference, that there is a gradual advance towards harmony between man's mental nature and the conditions of his existence. After finding that from it are deducible the various characteristics of Evolution, we finally draw from it a warrant for the belief, that Evolution can end only in the establishment of the greatest perfection and the most complete happiness (Spencer 1958: 510–11).

Then Spencer discussed dissolution, which is the process that takes place when evolution has run its course and the aggregate has reached "that equilibrium in which all its changes end" but is still subject to all the influences in its environment which may cause it to disintegrate. He described the process of dissolution of human societies, and of organisms, and then turned to the eventual fate of the unvierse. After the motion of the stars has become equilibrated and degraded to heat energy, we must expect a process of concentration of all the stars under the action of gravity. The stars will collide, coalesce, and then disperse again: "Action and reaction being equal, the momentum producing dispersion, must be as great as the momentum acquired by aggregation; and being spread over the same quantity of matter, must cause an equivalent distribution through space, whatever be the form of the matter" (Spencer 1958: 527). If the universe is finite and there is no loss of energy by radiation to the outside, there is some reason to believe that the cycle of condensation and diffusion can be continued indefinitely without loss. There would then be a sequence of "alternate eras of Evolution and Dissolution"; in other words, "there is suggested the conception of a past during which there have been successive Evolutions analogous to that which is now going on; and a future during which successive other such Evolutions may go on— ever the same in principle but never the same in concrete result" (Spencer 1958: 529).

What did the scientists think of Spencer's speculations? The reactions of two of them, John Tyndall and James Clerk Maxwell, were expressed in letters to Spencer in 1873 (Duncan 1908). Tyndall said that these chapters "were never satisfactory to me" and should be rewritten, but did not offer any specific criticisms. Maxwell thought that this type of speculation should be encouraged, though he refrained from giving his approval to any of Spencer's particular conclusions:

> Mathematicians by guiding their thoughts always along the same tracks, have converted the field of thought into a kind of railway system, and are apt to neglect cross-country speculations.
>
> It is very seldom that any man who tries to form a system can prevent his system from forming round him, and closing him in before he is forty. Hence the wisdom of putting in some ingredient to check crystallisation and keep the system in a colloidal condition. Candle-makers, I believe, use arsenic for this purpose. . . . But you seem to be able to retard the crystallisation of parts of your system without stopping the process of evolution of the whole, and I therefore attach much more importance to the general scheme than to particular statements (Duncan 1908: 162).

Spencer wanted Maxwell to confirm his notion that molecular motion is "rhythmic" but Maxwell insisted that insofar as rhythmic implies regularity and periodicity, this word is inappropriate; molecules move randomly, at least from the viewpoint of the human observer. Nevertheless it appears that Spencer eventually returned to his own ideas on most scientific subjects, and was not too disturbed by the refusal of scientists to approve them.

"La misérable race humaine périra par le froid."

"Ce sera la fin."

The Heat Death

Spencer's suggestion that dissipation produces a return to a balanced state is similar to an idea that Gustav Fechner, founder of psychophysics, elaborated in 1873. The "tendency toward stability," introduced by Fechner in order to develop a molecular theory of evolution, was later used by Sigmund Freud as a theoretical basis for his "death instinct," which was in turn interpreted as an application of the second law of thermodynamics by other writers.

The "heat death" was the concept, made explicit by Kelvin, Helmholtz and Clausius, of a final state of the universe that would result from the dissipation of all useful energy by transformation into heat at a uniform temperature. Naturally not everyone was satisfied with the pessimistic consequences that these physicists believe to be implied by the second law of thermodynamics, and there were many attempts to disprove or circumvent them. Some were grand schemes to reconcentrate the energy of the universe by reflecting energy at the boundaries of space (Rankine 1852), while others involved detailed arguments about molecular processes. The best-known plan of the latter type is the one involving "Maxwell's demon," described as "a being whose faculties are so sharpened that he can follow every molecule in its course" (Maxwell 1883: 328). The demon guards a sliding door between two containers filled with gas, one hotter than the other; he takes advantage of the fact that, according to Maxwell's velocity distribution law, there will be some molecules on the hot side with speeds lower than the average speed on the cold side, and some on the cold side with speeds higher than the average on the hot side. The demon allows only the faster molecules to pass from the cold to the hot side, and only the slower ones in the opposite direction, closing the frictionless door (an action that theoretically requires no work) in all other cases. He thus increases the average speed on the hot side and decreases it on the cold side, thereby in effect causing heat to flow from cold to hot, violating the second law of thermodynamics (Knott 1911; Maxwell 1871).

Maxwell himself thought that his scheme showed that the validity of the second law depends on the nonexistence of a demon who can sort out individual molecules; the law is nevertheless almost certainly valid in the real world, for all practical purposes. Balfour

Stewart and P. G. Tait (1875) suggested that the Maxwell demon might operate in an "unseen universe," linked by human thought to this one, so that we could look forward to a future state untroubled by dissipation of energy. Later scientists tried to identify Maxwell demons with enzymes in order to account for the apparent violations of the second law in biological systems.

Other efforts to avoid the consequences of the second law challenged Ludwig Boltzmann's attempt (1872) to derive what was later called the "H theorem"—an alleged proof that collisions among molecules will increase the entropy of a gas. Boltzmann used Newtonian mechanics to analyze molecular collisions, yet his result seems to contradict the reversibility of Newton's laws: entropy is in general larger after a collision than before.

There were two basic criticisms of Boltzmann's conclusion that molecular collisions lead to irreversible change, both of which appealed to the apparent inconsistency between that conclusion and Newtonian mechanics. The first criticism became known as the "reversibility paradox"; it was first discussed by Kelvin in 1874 but is generally attributed to Boltzmann's Viennese colleague Josef Loschmidt. The point is simply that in Newtonian mechanics, any particular sequence of motions of a system of particles can run equally well backwards or forwards. If the entropy increases for one sequence, it must decrease for the other, and there seems to be no justification in the laws of mechanics themselves for selecting the forward rather than the backward sequence. Whereas Kelvin argued that a gas with a very large number of molecules will nevertheless almost always appear to behave irreversibly, Loschmidt believed that the second law of thermodynamics does not imply irreversibility at all. He claimed that it could be formulated as a purely mechanical principle without reference to the sequence of events in time, in which case one could "destroy the terroristic nimbus of the second law, which has made it appear to be an annihilating principle for all living beings of the universe; and at the same time open up the comforting prospect that mankind is not dependent on mineral coal or the sun for transforming heat into work, but rather may have available forever an inexhaustible supply of transformable heat" (Loschmidt 1876: 135).

Boltzmann's reply (1877) to Loschmidt, while not a completely satisfactory solution of the reversibility paradox, was nevertheless a major step forward in theoretical physics, for it involved his first explicit statement of the famous relation between entropy and probability. The state of a system—that is, the observed macroscopic state, which may really be a collection of many conceivable molecular states—was assigned an entropy value proportional to the logarithm of its relative probability W. Roughly speaking, W is the number of molecular states corresponding to the observed macroscopic state. (This definition needs some modification when there is a continuum of possible molecular states.) A state which has low probability may be considered highly ordered, just as a bridge hand containing 13 cards of the same suit is highly ordered since there are only four possible hands of this kind. States with high probability may be considered disordered, like the mediocre kind of hand you ordinarily expect to get after shuffling the deck; there are many possible permutations and combinations which one would lump together as being equivalent.

According to Boltzmann's interpretation of entropy, the message of the second law of thermodynamics becomes: there is a universal tendency for things to get more and more disordered. The molecular explanation of this tendency is simply that the overwhelming majority of all possible states are disordered, and thus if you start out from an ordered state the likelihood that you will reach some disordered state after any finite period of time is very high. The answer to Loschmidt's reversibility paradox is that if you start from one of the handful of disordered states that happens to have evolved from an ordered state and reverse the velocities, you will indeed return to that ordered state, but if you start from any of the great number of other disordered states you will reach yet another disordered state by reversing the velocities. Thus the second law is no longer an absolute law of nature, but a statistical one that is obeyed with very high probability.

The other criticism of the H theorem is the "recurrence paradox," or eternal return. One of the major objectives of eighteenth-century mathematical physics had been to prove that the stability of the solar system is a consequence of Newton's laws of motion. By the beginning of the nineteenth century this goal had apparently been reached by the work of such scientists as Lagrange, Laplace, and Poisson. Consequently it was generally believed that the earth had remained at the same average distance from the sun for an indefinitely long time in the past, and that physical conditions on the surface of the earth had been roughly the same as they are now for countless millions of years. This assumption was the basis for uniformitarian geology (see Chapter III) but as Kelvin noted, the theory that predicted stability was only an approximate one. Henri Poincaré attempted to give a better proof of stability on the basis of an exact theory, and the result was his theorem stating that every mechanical system must almost certainly return infinitely many times as close as one wishes to its starting position. If this theorem is applicable to gases (and Boltzmann conceded that it is) then it appears to refute the H theorem, for it implies that the entropy cannot continually increase but must eventually decrease and return to its initial value.

Both the recurrence and the reversibility paradoxes were used (despite the opinions of Kelvin and Loschmidt) to attack the "mechanistic" viewpoint of the kinetic theory by showing that the theoretical consequences of mechanism are incompatible with the experience of irreversibility. The scientists of the neoromantic school known as energetics preferred to attribute absolute validity to the second law, rather than merely statistical validity which would be compatible with the kinetic theory.

Another possible solution of the dilemma, which presumably would have been satisfactory to the realists, would be to show (as Loschmidt thought) that the second law of thermodynamics can be reduced to purely mechanical principles. The problem of irreversibility and the directionality of time would thereby be avoided, or at least banished from physics to metaphysics. In any event, the project was considered so important that the British Association for the Advancement of Science appointed a committee to study it, and reports were presented at the 1891 and 1894 meetings of the Association.

The second of these meetings, held at Oxford, is of considerable historical interest. The famous debate between Bishop Wilberforce and T. H. Huxley had taken place in the same room (the Sheldonian Theatre) in 1860. Since that time the theory of evolution had become generally accepted, but bitter feelings were still harbored in some quarters. Lord Salisbury delivered the Presidential Address; Huxley agreed beforehand to second Lord Kelvin's vote of thanks, even though he knew that Salisbury intended to deprecate evolution in his speech. As Huxley rose to speak, there was a tremendous burst of applause from the audience; Huxley then "veiled an unmistakable and vigorous protest in the most gracious and dignified speech of thanks" (Peterson 1932: 298).

Boltzmann also attended this meeting, and participated in discussions with the British physicists on the foundations of the kinetic theory of gases. Shortly afterward he restated his views in a letter to *Nature*, taking Salisbury's view that "Nature is a mystery" as his starting point. He took this to mean that the Boscovich theory, which treats atoms as point centers of force, "is refuted almost in every branch of science" (Boltzmann 1895: 413). But he took great pains to dissociate the kinetic theory of gases from the Boscovich doctrine:

> We can hardly doubt that in gases certain entities, the number and size of which can roughly be determined, fly about pell-mell. Can it be seriously expected that they will behave exactly as aggregates of Newtonian centres of force, or as the rigid bodies of our Mechanics? (Boltzmann 1895: 414).

Boltzmann then gave his interpretation of the principle of dissipation of energy by postulating an "H curve" with certain properties. He believed it can be proved that if H (which is essentially the negative of the entropy) is not near its minimum equilibrium value, then it is probably very close to a relative maximum, so that it is almost certain to decrease if one looks either backward or forward in time. His reasoning seems rather dubious but it has been accepted by most physicists. He further suggested that the universe as a whole is always in a state of thermal equilibrium, but that if it is sufficiently large there must always be fluctuations which could produce, for example, the present state of our own world. The summits of the H curve, in general, "would represent the worlds where visible motion and life exist" (1895: 415).

While according to Boltzmann's interpretation the increase of entropy in natural processes is not inconsistent with a mechanistic view, and can to some extent be explained by it, he also admitted that mechanism alone is not sufficient for deducing rigorously a particular time-direction. The fact that entropy always increases in our own experience may have to be attributed to the circumstance that, as living beings, we represent an exceptional choice of initial conditions from the physical point of view; any world containing organic structures must have very low entropy already, so its entropy is unlikely to decrease any further.

Boltzmann suggested further (1897a) that the time-sense of an organism or a civilization is determined by the direction of increasing entropy, so that to say "entropy increases with time" is a mere tautology. (It is ironic that Boltzmann thus adopted the viewpoint of one of his critics, Ernst Mach, who made a similar suggestion in 1894.) This raises the possibility that elsewhere in our universe there may be worlds where the direction of time is opposite to our own. Boltzmann's speculation has been explored in some detail by the philosopher Hans Reichenbach (1956), and revived (without acknowledgment) by cosmologists in connection with the theory of the oscillating universe. In the contracting phase, time as perceived by the observer goes in the opposite direction, so he thinks his universe is expanding; thus the statement that the universe is expanding may also be a tautology if the direction of time is correlated with the expansion or contraction.

Another outcome of the nineteenth-century attempts to explain irreversibility was the suggestion that it may be necessary to postulate some kind of randomness at the molecular level. Kelvin stated bluntly in 1892 that "the fortuitous concourse of atoms is the sole foundation in Philosophy" on which the second law of thermodynamics can be based (1894: 464). S. H. Burbury (1894) pointed out that Boltzmann had in effect made such an assumption in his original proof of the H theorem, and Boltzmann admitted that a postulate of "molecular disorder" could be used in this proof though he did not think the postulate was always valid in nature.

Max Planck, in his researches on radiation theory before 1900, proposed to show that irreversibility is a consequence of the interaction between matter and radiation as described by electromagnetic theory. He was skeptical of the value of kinetic theory and atomistic hypotheses, preferring instead to work as far as possible with general principles such as the laws of thermodynamics. But Boltzmann (1897b, 1898) showed that Planck's attempt to derive irreversibility from electromagnetic theory was incorrect, since the latter theory is based on time-reversible equations just as is Newtonian mechanics. Planck was thus forced to postulate that the radiation which interacts with matter is always "natural"—the relation between different frequency components is random or indeterminate—and he recognized that this postulate is essentially the same as the Burbury-Boltzmann "molecular disorder" assumption for gases.

Thus by the end of the nineteenth century it appeared that the second law of thermodynamics could not be completely explained without introducing some kind of statistical assumption which might involve randomness at the molecular level. Such an assumption would have to go beyond what is sometimes called the "statistical interpretation of thermodynamics," i.e., the viewpoint that while molecular motions are themselves determined "in principle" by mechanical laws, one has to describe them statistically because of the difficulty of computing with the positions and velocities of large numbers of molecules. According to that viewpoint the use of statistics is merely a matter of convenience. In the twentieth century, statistics became more and more convenient as the concept of determinism became less and less meaningful. The efforts of Maxwell, Boltzmann, Kelvin, Burbury, and Planck to explain the second law of thermodynamics were still a topic of lively debate in the early 1900s, and probably played some role in the overthrow of determinism that came with quantum mechanics in the 1920s (Brush 1976).

The notion that history repeats itself—that there is no progress or decay in the long run, but only a cycle of development that always returns to its starting point—has been inherited from ancient philosophy and primitive religion. As Shelley expressed it, in his poem *Hellas* (1822):

> The World's great age begins anew,
> The golden years return,
> The earth doth like a snake renew
> Her winter weeds outworn . . .

> Another Athens shall arise,
> And to remoter time
> Bequeath, like sunset to the skies
> The splendour of its prime . . .

But this recurrence is not entirely desirable, as Shelley indicates at the end of the poem:

> Oh, cease! must hate and death return?
> Cease! must men kill and die?
> Cease! drain not to its dregs the urn
> Of bitter prophesy.
> The world is weary of the past,
> Oh might it die or rest at last!
> —(Shelley 1965: 52–53)

It has been pointed out by some scholars that belief in recurrence, as opposed to unending progress, is intimately connected with man's view of his place in the universe, as well as with his concept of history. Starting, in most cases, from a pessimistic view of the present and immediate future, it denies the reality or validity of human actions and historical events by themselves; actions and events are real only insofar as they can be understood as the working out of timeless archetypal patterns of behavior in the mythology of the society. This attitude is said to be illustrated in classical Greek and Roman art and literature, where there is no consciousness of past or future, but only of eternal principles and values. By contrast the modern Western view, as a result of the influence of Christianity, is deeply conscious of history as progress toward a goal. (Momigliano [1966] dissents from this interpretation.) Nevertheless, the cyclical view has by no means died out, and can easily be recognized in the persistent tendency to draw historical analogies and comparisons.

The suggestion that eternal recurrence might be proved as a theorem of physics, rather than as a religious or philosophical doctrine, originated at about the same time from the French mathematician Henri Poincaré and the German philosopher Friedrich Nietzsche. Nietzsche encountered the idea of recurrence in his

The Heat Death

studies of classical philology, and again in a book by Heine. It was not until 1881 that he began to take it seriously; then he devoted several years to studying physics in order to find a scientific foundation for it. He read Vogt's *Die Kraft* (one of the materialistic popularizations of science), preferred the theories of Boscovich to those of Robert Mayer, sided with the school of energetics against the atomistic-mechanical doctrines, and knew of Kelvin's work through the expositions of Otto Caspari and Zoellner.

The results of Nietzsche's thinking were expressed most completely in the last section of his book *The Will to Power,* published posthumously in 1901–4 but written during the years 1884–88. He began by noting a "will for ruin" in Europe and a general pessimistic attitude; while this may be useful for the purpose of eliminating degenerate races that deserve to perish, "general levelling down to mediocrity must be avoided." Thus Nietzsche, like many other writers, predicted eternal recurrence in the face of an apparent degeneration and disintegration of civilization, an apparent decline from a previous "Golden Age"; his prediction was intended to stimulate a revival of those heroic qualities that will characterize the new Golden Age in the future.

Nietzsche's "proof" of the necessity of eternal recurrence was as follows: "If the universe had a goal, that goal would have been reached by now" since the universe, he thought, has always existed; the concept of a world "created" at some finite time in the past was stigmatized as a meaningless relic of the superstitious ages. He absolutely rejected the idea of a "final state" of the universe, and further remarked that "if, for instance, materialism cannot consistently escape the conclusion of a final state, which William Thomson has traced out for it, then materialism is thereby refuted." He continued:

> If the universe may be conceived as a definite quantity of energy, as a definite number of centres of energy—and every other concept remains indefinite and therefore useless—it follows therefrom that the universe must go through a calculable number of combinations in the great game of chance which constitutes its existence. In infinity, at some moment or other, every possible combination must once have been realized; not

only this, but it must have been realized an infinite number of times. And inasmuch as between every one of these combinations and its next recurrence every other possible combination would necessarily have been undergone, and since every one of these combinations would determine the whole series in the same order, a circular movement of absolutely identical series is thus demonstrated: the universe is thus shown to be a circular movement which has already repeated itself an infinite number of times, and which plays its game for all eternity. This conception is not simply materialistic; for if it were this, it would not involve an infinite recurrence of identical cases, but a final state. Owing to the fact that the universe has not reached this final state, materialism shows itself to be but an imperfect and provisional hypothesis (Nietzsche 1964: 90).

Nietzsche was mistaken in thinking that his conclusion is not materialistic; on the contrary, it is precisely the materialistic doctrine that the world consists of nothing but matter and energy, evolving through a fixed sequence of states determined by the laws of classical mechanics—it is precisely this doctrine that has recurrence as its inevitable consequence. Nietzsche's own argument, though it makes no pretense of mathematical rigor, is one of the clearest and most persuasive statements of the "recurrence paradox" published by anyone in the nineteenth century, and it is all the more surprising to find it in the works of a writer who is best known for his profusion of mystifying and frequently hysterical aphorisms, as in *Also Sprach Zarathustra*.* He merely made the very common error of identifying Kelvin and the principle of dissipation of energy with materialism. In the historical context of late nineteenth-century physics (since it was the recurrence paradox that was used to attack materialism, on the grounds that recurrence is an inevitable consequence of the kinetic theory of gases, and contradicts the second law of thermodynamics), we must conclude that the effect of Nietzsche's argument is just the opposite of what he thought it should be. If there *is* eternal recurrence, so that the

---

*Before crediting Nietzsche with too much originality one should compare the discussion of Vogt (1878, pp. 88–90) which he is known to have read.

second law of thermodynamics cannot always be valid, then the materialist view would be substantiated.*

Poincaré, who did give a mathematical proof of the recurrence theorem (1890), was also interested in its philosophical implications. He pointed out in 1893 that the theorem was one of the difficulties encountered by scientists who accepted the mechanistic conception of the universe. Mechanism implies that all phenomena must be reversible, yet experience shows that many irreversible phenomena exist in nature. To escape the contradiction, physicists have postulated "hidden movements": for example, if we didn't know that the earth rotates, we would regard the motion of the Foucault pendulum as "irreversible," but having discovered that the earth does rotate, we can *imagine* that it might just as well be rotating in the opposite direction. Hence we don't consider this a contradiction of the principle of reversibility. Similarly one might suppose that there are motions in the molecular world which account for microscopic irreversibility, and which are "in principle" reversible.

Poincaré alluded briefly to Maxwell's demon, and the argument that "the apparent irreversibility of natural phenomena is . . . due to the fact that the molecules are too small and too numerous for our gross senses to deal with them" (1893: 536). Yet, while the kinetic theory of gases based on this premise is, according to Poincaré, "up to now the most serious attempt to reconcile mechanism and experience" (ibid.), it still has not overcome the difficulties; his recurrence theorem, which would seem to apply to the entire world if the kinetic theory is valid, contradicts the "heat death" theory. If one attributed absolute validity to the second law, then the universe, instead of returning to its initial state, would tend toward a final state of uniform temperature.

One could reconcile the two theories, he suggested, by assuming that the heat death is not permanent but only lasts a very long time, so that the universe, after slumbering for millions of millions of centuries, will eventually reawaken. Then, as Poincaré put it, "to see heat pass from a cold body to a warm one, it will not be neces-

---

*Thus the eternal return theory was enthusiastically supported by the socialist agitator Blanqui, writing from his prison cell; see Blanqui (1872) and the editor's note appended to this article.

sary to have the acute vision, the intelligence, and the dexterity of Maxwell's demon; it will suffice to have a little patience" (1893: 536).

The mathematician Ernst Zermelo insisted that the recurrence paradox was a fatal objection not only to the kinetic theory of gases but to the mechanical view of nature in general. Any mechanistic theory must predict recurrence, contrary to the dissipation principle. He dismissed Boltzmann's reply (1896) on the grounds that the properties of the postulated H curve were mathematically impossible (Zermelo 1896). It was in reply to this criticism that Boltzmann suggested that the direction of time may fluctuate with the direction of entropy change, thus making the second law true by definition rather than by deduction from kinetic theory.

Writers on statistical mechanics and kinetic theory usually present the recurrence paradox as a purely technical curiosity without mentioning its importance for the study of the relations between science and culture. At the same time, scholarly critics of Nietzsche often seem to be unaware that his argument for recurrence is not at all nonsense. The situation is further confused by the fact that Boltzmann, while accepting recurrence as a mathematical consequence of the laws of motion, was mainly concerned with the task of convincing physicists that it could have no empirical significance because the recurrence time would be enormously long. The dispute between Boltzmann and his energetist opponents was therefore reduced to the question whether the atomic-mechanistic theories should be thrown out because they permitted possible deviations from the second law with infinitesimal probability rather than excluding them entirely. This was not a good battleground for waging the war of materialism vs. idealism, and it is hardly surprising that it failed to sustain the interest of the scientific public for very long.

Chapter VI

Realism and Neoromanticism

> Physical science, cultivated exclusively for its own sake and as an end, inevitably tends, in our poor nature, to contract, to carnalise, and, in some cases, even to embrutify, if we may so speak, the judgment of its devotees: the mind becomes insensibly impregnated with the muddy qualities of the evil through which it rises, so that when applied to questions of human life, it becomes confused, giving birth to a monstrous flood of misshapen revelations.
> —(*British and Foreign Evangelical Review*, quoted by Young 1871: 344)

The early 1870s were good years for science. Not spectacular years, like 1543 or 1905, when revolutionary theories were published, but a time when scientists could proudly observe the consolidation of major achievements of preceding decades. James Clerk Maxwell's *Treatise on Electricity and Magnetism* (1873) provided an impressive synthesis of the discoveries and concepts of Oersted, Ampère, Faraday, and Kelvin, capped by his own interpretation of the physical nature of light, and the prediction that electromagnetic waves could exist at any frequency. The molecular-kinetic theory of matter, founded on the ideas of Clausius and Maxwell developed in the late 1850s, was cast into powerful and useful forms in major works of Ludwig Boltzmann (1872) and Johannes Diderik van der Waals (1873). And the biological sciences had at last found a persuasive explanatory scheme: Charles Darwin's *Descent of Man* (1871) provided the inevitable application of his theory of evolution by natural selection to the crucial problem of how the human body and mind have developed to their present state. Before the end of the next decade Boltzmann (1886) was to predict that the nineteenth would be known as "Darwin's Century."

The self-confidence furnished by such triumphs prompted a few Victorian scientists to issue a bold challenge. To those religious persons who believed in the effectiveness of prayer, they proposed a crucial experiment: let's see if the prayers of an entire country, directed toward a single desired outcome, can yield a measurable effect. Thus began the "prayer-test" debate of 1872–73, a remarkable but long-forgotten skirmish in the centuries-old warfare between science and religion.

The first shot was fired by John Tyndall, known in scientific circles for his researches on radiant heat, and to a larger public as an exponent of "scientific materialism," the view that all natural phenomena can eventually be explained by the laws of physics and chemistry. Tyndall sent to the *Contemporary Review*, with his own brief note of introduction, an anonymous article later attributed to a London surgeon, Sir Henry Thompson. As Tyndall said, the ostensible purpose of the article, entitled "The 'Prayer for the Sick,' " was "to confer quantitative precision on the action of the supernatural in Nature" (Tyndall 1872a: 206). He made little attempt to conceal his hope that the clerics who claim "the habitual intrusion of supernatural power in answer to the petitions of men" (1872a:205) could now be made to put up or shut up.

The author of "The 'Prayer for the Sick' "asserted that if we are "in contact with a source of power available for human ends (or affirmed to be so on high authority)" we have a duty "to estimate its value" ([Thompson] 1872: 206). (Quantitative determination of the energy value of the various forces of nature had been a fruitful scientific activity since the experiments of James Prescott Joule and others in the 1840s.) Of the several purposes for which prayer is recommended, there is one whose consequences can now be objectively evaluated: the prayer that particular persons recover from sickness. Why should we not test the efficacy of prayer in the same way we test that of any other proposed remedy for a disease: select a group of patients suffering from the disease, administer the remedy in carefully measured amounts, and observe the effects. The crucial feature of any such test is not only that we have accumulated evidence on what happens to patients who have *not* been given this remedy, but that we set up a "control group" of patients suffering from the disease, this group being as similar as possible to the experimental group in all respects, and subjected to exactly the same conditions for the same period of time, except that its members are not given the remedy being tested.

The test, then, would be to designate a particular hospital containing patients with diseases whose mortality rates are well established by past experience, and recommend that these patients be made "the object of special prayer by the whole body of the faithful" for a period of not less than three to five years. At the end of this time, the mortality rates in this group would be compared with those suffering from the same diseases in other hospitals "similarly well managed." Those who really believe that they are not wasting their time praying to God should welcome such an opportunity of "demonstrating to the faithless an imperishable record of the real power of prayer" ([Thompson] 1872: 210).

That was certainly not the response of the representatives of organized religion who published their comments on the proposed test in the following months. An editorial writer for the *Spectator* called it "revolting to the spirit of Christian prayer," and professed to approach the subject only with "reluctance and disgust" (Means 1876: 28,25). As it turned out he could not resist the opportunity to flail the "arrogant physicists" for their contempt of religion (1876:23). (At that time the term "physicist" was used in the original sense, meaning a student of nature in general, and included physiologists and geologists as well as men like Tyndall who were physicists in the modern sense.) He pointed out that God is hardly likely to cooperate in an experiment whose real purpose is not to heal the sick but to provide a "scientific" measure of His power. In fact the Biblical admonition "Thou shalt not tempt the Lord thy God" would seem to be specifically designed for this situation (Tyndall's anonymous friend being thus cast in the role of the Devil) (Means 1876: 28).

The Reverend Richard Frederick Littledale, in an article on "The Rationale of Prayer" published in the *Contemporary Review*, promised to discuss the prayer-test but instead devoted most of his space to a lengthy attack on Tyndall's "crusade against prayer," conducted in earlier speeches and articles. He went on to denounce the whole tribe of physicists who "seem unable to rise out of the plane of material conceptions into broad moral and spiritual views, or even to look at phenomena belonging to other spheres of knowledge with scientific eyes" (1872:436–37). The longevity of Christianity, and the persistence of the practice of prayer, are "facts" about the world which narrow-minded physicists choose to ignore, he complained. But Rev. Littledale at last succumbed to the idea of a "scientific test" of the value of prayer. Instead of the one proposed by Tyndall's friend, he suggested that "a tabular comparison of the results severally attained by nurses who work for God"—such as the "sisters of conventual societies, who are moved by piety by their labor of love, and sustained in it by prayer"—and "nurses who work for money" should be made (1872:450). (There is no indication that such a test was ever made, or that theologians' belief in the power of prayer would have been affected in any way by its outcome.)

Reinforcement for the physicists soon arrived from another quarter. Francis Galton, already well known as the author of *Hereditary Genius* and *English Men of Science*, revealed that he had for several years been collecting statistical data on the efficacy of prayer. Tables of the average life span of various classes of persons show that kings and queens, who are usually the object of public prayer by their subjects, die earlier than lawyers, gentry, and military officers. Members of the clergy, presumably a prayerful class of men, do not live significantly longer than lawyers and physicians. Missionaries, whose effectiveness in spreading the gospel is crucially dependent on their living as long as possible after learning the language and habits of the country to which they are sent, frequently die shortly after arrival in spite of the many prayers that accompany them. The proportion of stillbirths suffered by praying and non-praying classes of parents appears to be identical, though there can be no doubt that if a person ever prays at all it will certainly be for the health of an expected child.

Tyndall came out fighting at the beginning of the next round. He recalled some of the points on which religion had previously given way to science: the existence of the antipodes, the motion of the earth, the age of the world, and the theory of evolution. Abandoning belief in the physical value of prayer might well be the next "act of purification" by which religion would free itself from dependence on superstition, now that science had developed methods for analyzing forms of energy, and could thus examine the claim that prayer "produces the precise effects caused by physical energy in the ordinary course of things" (1872b: 764). Tyndall did not want to be accused of flatly denying the value of prayer (though no one was in much doubt as to his opinion); instead he insisted that he was willing to admit "the theory that the system of nature is under the control of a Being who changes phenomena in compliance with the prayers of men" as a "perfectly legitimate" theory (1872b: 765–66), provided that it was to be considered subject to experimental test like any theory in science. Just as Newton's theory of light was abandoned when his prediction that light travels faster in water than in air turned out to be wrong, so the theologians should be willing to agree to a crucial experiment on the value of prayer and to abandon their theory if the result is unfavorable. Yet the theologians seem to resent the suggestion of such a test, either because they enjoy the very act of praying regardless of the results, or because they are still under the sway of medieval mysticism.

James M'Cosh, a Scottish theologian who had recently gone to America to become President of Princeton College, now entered the battle. He criticized Tyndall for confusing the methods of the physical sciences with those appropriate to religion and moral philosophy, and for misconstruing the type of "answer" God may give to prayers. The notion of a "control group" of sick persons, for whom one deliberately does *not* pray, is so repugnant to the true Christian as to cast doubt on the sanity of the enterprise. What if a skeptical young man, instructed by his father to be virtuous in order to enjoy ultimate happiness, were to propose an experiment with the boys of a poorhouse, "one-half of whom are allowed every indulgence, while the other half are exposed to restraint"— would that be considered a reasonable "test" of moral philosophy, or would the father be justified in rejecting it with the assertion "that virtue is a thing binding on us, that by its very nature . . . is fitted to lead to happiness, and by pointing to the issues of virtue and vice seen in common life"? (M'Cosh 1872: 780).

As for the "effectiveness" of prayer, M'Cosh noted that there are many cases in which (as *later* becomes evident) it is the wisdom of God to answer the prayer by *denying* what the petitioner *thinks* he wants in order to lead him on a better path. There are enough of these cases to show that one cannot simply tabulate the results of prayers as "answered" or "not answered." For example, when Prince Albert was sick with a raging fever a few years ago, hundreds of thousands prayed for his recovery, apparently to no avail. But shortly after his death, Queen Victoria's advisers urged her to declare war on America (this is M'Cosh's account "on what I believe to be good authority"). The Queen refused "because her departed husband was always opposed to such a fratricidal proceeding" (M'Cosh 1872:779). Yet one might suppose that if the Prince had been still alive his influence would not have been strong enough to stop the war. So in refusing to follow the wishes of those who prayed for Albert's recovery, God was really acting in their best interests. In the same way, for reasons we cannot now imagine, God might refuse to give preference to the patients in the hospital ward for which everyone is praying.

Of course it was just this kind of haggling about individual cases that Galton wanted to avoid when he proposed to treat the whole question statistically. Throughout the course of the debate, which continued in the *Spectator* and *Contemporary Review* for several months, hardly anyone attempted to refute his evidence, though other examples were brought forward in which the prayers of large numbers of people over long periods of time allegedly *had* been effective—for example, in promoting the spread of Christianity and the longevity of the Papacy. A more effective tactic was to admit that the primary purpose of prayer is not to request specific physical actions but rather to gain spiritual strength that may be employed in ways that may have little to do with the subject of the prayer. This viewpoint would make Galton's statistics irrelevant, and set aside the possibility of any kind of scientific test of the efficacy of prayer. Yet the theologians of the 1870s were reluctant to stick to this line; it would have looked too much like a retreat, in the eyes of congregations accustomed to being urged to pray for specific acts of divine providence.

The "physicists" came out of this debate with their self-confidence intact, even though they do not seem to have deprived anyone of his faith in the value of prayer. The theologians effectively pressed the argument that religious beliefs cannot be tested by scientific experiment, though at some cost to their status in a century that was according increasing prestige to the scientific world view. At least their position was more tenable than that of the spiritualists who did perform experiments on psychic phenomena but claimed that it was in the nature of these phenomena to disappear when a skeptical observer was in the room.

Perhaps the net effect of the debate was merely to widen the chasm between the scientific and religious viewpoints. No one pointed out that the "crucial experiment" is almost as rare in science as it is in religion; for example, contrary to Tyndall's statement, physicists had already abandoned Newton's theory of light for a combination of other reasons two decades before the experiment on the speed of light in water. One does not have to accept the extreme conclusion that some observers have drawn from Thomas Kuhn's theory of scientific revolutions—that changing from one scientific paradigm to another is like a religious conversion experience—to realize that Tyndall's view of the role of experiments in theory-testing is unrealistic. A scientist always interprets experimental data within some theoretical framework, and necessarily hesitates to abandon that framework without compelling reasons going beyond the mere numerical discordance of a few observations. Conversely, people do change their religious beliefs, in part because of personal experience, and such behavior is not qualitatively different from that of a scientist who changes his theory as he acquires new experimental data.

The prayer test was finally performed, nearly a century later, by C. R. B. Joyce and R. M. C. Welldon (1965). They studied the effects of prayer on patients "suffering from chronic stationary or progressively deteriorating psychological or rheumatic disease" using the experimental method proposed by nineteenth-century scientists. They selected 38 patients matched in 19 pairs "as closely as possible for sex, age, and primary clinical diagnosis" and (in more than half of the pairs) for marital status and religious faith. One patient in each pair was prayed for, the other (as a "control") was not. The physicians who treated these patients were asked to evaluate their clinical state at the beginning and end of a trial period (from 8 to 18 months), not knowing which patient was being prayed for. The patients themselves were not aware of the experiment.

The results were somewhat inconclusive. Out of 12 pairs of patients for which it was definitely established that one patient did better than the other, the prayed-for patient showed greater improvement in 7 cases, the control patient in 5. By itself this result is not statistically significant. But the sequence of individual results suggested that prayer helped patients whose clinical state was evaluated after a short time, while it hindered those who for various reasons were not evaluated until several months after the end of the original trial period. Since "it was not known" whether the prayer groups continued their efforts after this period, one does not know how to interpret the fact that in each of the first six pairs for which the evaluation was completed the prayed-for patient did better, while in five of the remaining six the control patient did better.

According to a persistent myth invented by Robert A. Millikan (1927), physicists at the end of the nineteenth century regarded their subject as essentially complete; all the important discoveries were thought to have been made, and there remained only the determination of certain physical constants to a few more decimal places. This supposed mood of complacency was then disrupted by the quantum theory and relativity, just as, in the nostalgic view of history, Victorian serenity was disrupted by World War I.

Anyone who takes the trouble to read the writings of physicists and their critics in the 1890s should immediately recognize the falsity of the myth. It is true that one can find a number of indications of confidence in the power and success of science in the 1870s, and occasional statements by British and American physicists to the effect that all the great discoveries had already been made, but only a few of them continued to take this notion seriously in the 1890s (Schuster 1918, Badash 1972). The remark that future progress would depend on very accurate measurements had become a truism for men like A. A. Michelson, though they tended to attribute it to some great predecessor and follow it with a disclaimer (Millikan 1950:23). By 1894 even Michelson must have been aware that the Newtonian program of mechanical explanation had broken down, and that many of the leaders of science and philosophy had rejected not only mechanism but also belief in the existence of atoms. There was much talk of an urgent need for finding new foundations of physics, and some intellectuals pointed to the "bankruptcy of science." As Barbara Tuchman (1966) has noted, the nineteenth century appears calm and peaceful only to those who look back on it from the vantage point of the twentieth.

Boltzmann, in the second volume of his *Lectures on Gas Theory* (1898), protested the attacks on his kinetic theory and called himself "an individual struggling weakly against the stream of time" (Boltzmann 1964: 216). He seemed to be the only remaining defender of atomism, and felt it necessary to stipulate that he did not believe in the literal accuracy of the various atomic models which he had used in developing the kinetic theory. Max Planck, who stood with Boltzmann against energetics, did not yet have much enthusiasm for Boltzmann's statistical theories though he was later to use them with great effect in developing quantum theory. The most "advanced" and "sophisticated" theories were those that took a purely phenomenological viewpoint: scientific theories should deal only with the relations of observable quantities and should strive for economy of thought, rather than trying to explain phenomena in terms of unobservable entities. This attitude was in part based on the failure of one of the great projects of nineteenth-century physics: to construct a mechanical or atomic model of matter and of the ether which would enable one to explain thermal and electromagnetic properties by means of Newtonian mechanics. But it was also in part a reflection of the reaction against a materialism which threatened to become too successful in reducing life itself to matter and motion.

Before about 1870, there was little or no connection between the research that scientists were doing on atomic theories such as the kinetic theory of gases, and the controversies among philosophers about the nature of matter, even though both were apparently discussing the properties and motions of atoms. The atom of the chemists and physicists, which was employed to express quantitative relationships between observable properties of matter, was entirely distinct from the atom of the philosophers, which symbolized a doctrine about the real nature of matter. As long as both atoms remained entirely hypothetical and unobservable, there did not need to be any conflict between them. But when Josef Loschmidt (1865), Kelvin (1870), and others started to estimate numerical values for the size, mass, and charge of atoms, as though atoms really existed and might some day be observed through a microscope, the situation changed. Many sensitive intellectuals were becoming alarmed at the whole tendency of physical and biological science since the middle of the century. The idea had become prevalent that science would soon be able to provide a mechanistic explanation of all the phenomena in the universe in terms of the motions of atoms and the ether. At the same time it was being asserted that everything that was *not* susceptible to mechanical explanation could never be understood at all. This combination of attitudes can hardly be called a philosophical system, yet philosophers have expended prodigious efforts trying to refute it. It is most often labeled "materialism," although the multitude of meanings which have been attached to this word make it practically useless except as an epithet. Hardly anyone will admit to being a materialist himself, and therefore most definitions have been proposed by writers who were trying to describe a view which they detested.

In the usual classification scheme, "idealism" is the opposite of materialism; the former holds that only mental concepts and things of the spirit are real, whereas the latter holds that only matter is real. Idealism and materialism are the philosophical components of the movements that we call romanticism (or neoromanticism) and realism, respectively. The corresponding scientific viewpoints, which might be called positivism (or empirio-criticism) and scientific materialism, are not so much concerned with what is "real" as with what can be employed to construct a useful scientific theory.

While Susan Stebbing complains that "most physicists who have attempted to construct a philosophy upon the basis of physical researches have ended by elaborating some form of idealism" (1958: 265–66), Philipp Frank (1937) points out that professional philosophers have always tended to interpret a transition from a mechanical to a more formal mathematical theory as a movement toward idealism. It has been claimed, for example, that the use of abstract mathematical representations in quantum mechanics and relativity indicates that it is mental concepts rather than atoms which are being invested with reality.

To clarify our definitions of terms, we must note also the popular usages of the terms materialism and idealism as exemplified in the following statement of Theodore Roosevelt:

> Surely all of us ought to realize the need in this country of a loftier idealism than we have had in the past; and the further and even greater need that we should in actual practice live up to the ideals we profess. The things of the body have a rightful place and a great place. But the things of the soul should have an even greater place. Materialism, in the end, eats like an acid into all the finer qualities of our souls (Roosevelt 1960).

It should be evident that materialism and idealism as used here have almost nothing to do with the philosophical terms defined above; nevertheless, the popular connotations tend to confuse discussions of the philosophical ideas. Perhaps we should use the label "*crass* materialism" for the attitude that only the possession of material wealth and the enjoyment of sensual pleasure are of value in life, and "*fine* idealism" for the opposite belief that only adherence to ideals and concern for things of the soul are important.

The disgust for *crass* materialism was due partly to the mechanization of work and life in general that accompanied the rise of industrialism in the nineteenth century, together with the loss of faith in traditional religion. Some of this disgust was transferred to *scientific* materialism, especially in the biological sciences. Charles Lindbergh wrote in 1948:

> I grew up as a disciple of science. I know its fascination. I have felt that godlike power man derives from his machines. . . . Now I have lived to experience the early results of scientific materialism. I have watched men turn into human cogs in the factories they believed would enrich their lives. I have watched pride in workmanship leave and human character

decline as efficiency of production lines increased . . . (Packard 1960: 318).

It is rare to find anyone except a Marxist explicitly defending the materialist viewpoint, but here is one example from a book by the astronomer Fred Hoyle (1956):

> What is a materialist? In the popular view I suppose a materialist is a pretty unpleasant person who gobbles babies for breakfast. This is a view I do not agree with. I am a materialist and I haven't gobbled any babies, yet. . . . The essence of materialism lies in a refusal to separate Man and his environment into the mutually exclusive categories of "spiritual" and "material." Man is regarded as belonging to the Universe, not necessarily insignificantly, as a star or a galaxy belongs to the Universe. Star, galaxy, man, are all expressions of the structure of the Universe. No attempt is made to introduce the notions of value or importance. . . . The materialist . . . will only secure a complete victory over his opponents if he is able to show that the behavior of Man can indeed be understood with precision (Hoyle 1956: xix–xx).

The kinetic theory of gases may be considered the outstanding representative of materialism in physics, and its ups and downs are strongly correlated with trends of opinion in biology, geology, politics, and the arts. It is probably futile to try to determine any rigorous cause-and-effect relationship when interactions among human activities are involved; nevertheless it is tempting to suggest that the success of the mechanistic world view introduced by seventeenth-century physics encouraged materialistic explanations in the natural sciences, and that when biological theorizing was carried to such an extreme that it offended public opinion, the "backlash" against materialism was so strong that it also affected physics. (As noted above, the word "physics" was still occasionally used in a sense that included biology at this time.) Regardless of how much truth there may be in this, it is undeniable that the theory of natural selection—which purported to provide a mechanism by which the evolution of man from the simplest forms of life, and perhaps ultimately from inanimate matter, might be explained—was the most important scientific theory in the nineteenth century from the viewpoint of the nonscientific public. It certainly aroused much stronger feelings than the viscosity of gases, or the frequency spectrum of black-body radiation, or the law of combining proportions, and therefore it seems that one cannot properly study the history of any part of nineteenth-century science without paying some attention to the possibility of interactions with biology.

While popular scientific materialism claimed to encompass all nature and reduce spirit to matter in motion, the pronouncements of scientists who were regarded as materialists display considerable reluctance to bite off such an indecently large mouthful. On the contrary, they emphasize our ignorance of everything except a few properties of dead matter, and assert that the methods of science are unable to deal with much else except these properties. The essence of their materialism is in their contention that where mechanistic science fails, nothing else can succeed. Thus not atheism but agnosticism was the characteristic stance of these scientists (the word itself was first used in the modern sense by T. H. Huxley in 1869). A famous statement of this position is contained in the 1872 lecture of the physiologist Emil du Bois-Reymond. Du Bois-Reymond maintained that the only true and exact science is mechanics; all other modes of investigation based on qualitative principles (moral, aesthetic, etc.) can never lead to reliable results. Hence the limits of our possible knowledge of the world are determined by the extent to which purely mechanical principles can be applied. As for everything else, not only do we know nothing, but we can never hope to know anything. The lecture ends with the slogan: "Ignorabimus" (DuBois-Reymond 1874).

In England, two of the most influential popularizers of science, John Tyndall and Thomas Huxley, were bold enough to risk the onus of materialism in lectures delivered in 1868. Huxley predicted that physiology will eventually "extend the realm of matter and law until it is coextensive with knowledge, with feeling, and with action.... The consciousness of this great truth weighs like a nightmare, I believe, upon many of the best minds of these days.

They watch what they conceive to be the progress of materialism, in such fear and powerless anger as a savage feels when, during an eclipse, the great shadow creeps over the face of the sun. The advancing tide of matter threatens to drown their souls; the tightening grasp of law impedes their freedom; they are alarmed lest man's moral nature be debased by the increase of his wisdom" (1948: 21). Tyndall insisted that mechanical laws apply even to the growth of a grain of corn:

> Given the grain and its environment, with their respective forces, the purely human intellect might, if sufficiently expanded, trace out *a priori* every step of the process of growth, and by the application of purely mechanical principles, demonstrate that the cycle must end, as it is seen to end, in the reproduction of forms like that with which it began. A necessity rules here, similar to that which rules the planets in their circuits round the sun (1897: 84).

Even a scientist with clerical affiliations might be caught up in the wave of enthusiasm for extending physical theories to all phenomena. The Reverend John Hewett Jellett, in an address to the British Association in 1874, predicted that eventually every science would be subject to mathematical laws of mechanics; moreover, "Let no one presume to fix the bounds of science" (1874: 323).

The realist movement believed that materialistic explanations were the ultimate goal of science, though few would state their opinions as explicitly as Tyndall, Huxley, and Jellett. The "reaction against materialism" was therefore one of the major themes of the neoromantic movement.

Just as the transition from romanticism to realism can be followed under various names and aspects just before the middle of the nineteenth century, so we can identify a transition from realism to neoromanticism in the arts and sciences in the late nineteenth century. We have already discussed in detail one component of this transition: the idea of dissipation of energy; another component, the reaction against materialism, will be described in the last part of this chapter.

One characteristic of neoromanticism is an insistence on the uniqueness and autonomy of each academic discipline or mode of artistic expression. Symbolism, impressionism, and aestheticism all rejected the demand that art should provide an accurate description of nature and life, and emphasized the right of the artist to follow his own inclinations. This viewpoint was expressed by Oscar Wilde in his American lectures (published in 1906), with reference to the painting of Burne-Jones and the poetry of Morris, Rossetti, and Swinburne. Benedetto Croce, the leader of neo-idealism in historical writing, declared that history must become independent of the natural sciences; it should be scientific only in the sense that it followed its own rigorous methods in handling data. Sociology, under the leadership of Weber and Durkheim, also refused to admit the validity of analogies and models derived from the natural sciences, and established itself as an independent discipline with its own methods. Psychology outgrew the primitive psychophysics of Fechner, and began to accept data acquired by introspection and hypnosis; Gestalt psychology and psychoanalysis originated in this period, along with such diverse theories as nativism, "content" psychology, "act" psychology, and "imageless thought." Anthropology, following Franz Boas, renounced evolutionary explanations of primitive cultures, and until quite recently accepted Durkheim's thesis that there are "social facts" separate from psychological and biological facts (Freeman 1966).

As the rational humanism of the realist period declined, political leftism subsided. Midcentury cosmopolitan liberalism was thought to have gone too far in breaking down the cohesiveness of traditional society. Conservative and patriotic movements (French Legitimists, British Imperialists, Spanish Carlists, Portuguese Miguelists, Prussian Junkers) arose and gained power in many countries. Even the intellectuals began to lose faith in democracy; Pareto scoffed at the possibility of popular government and saw political history as a cyclic alternation of control by elite groups. Nationalism and *Realpolitik*, whose roots may be found to lie in the realist period, now developed into imperialism, the "divine right of expansion," and "manifest destiny."

The phrase "new romanticism" was introduced by the German publisher Eugen Diederichs and was associated with the "Volkisch" movement that ultimately provided the background for Nazism. In modern terms that movement was racist and sexist, exalting the superiority of the white Nordic male. By encouraging this cultural reaction against materialism, German intellectuals in the late nineteenth and early twentieth centuries ultimately brought the entire romantic movement into disrepute among post-1933 historians (Mosse 1964, Ringer 1969, Gasman 1970).

Other aspects of neoromanticism were the revival of mysticism and spiritualism in religion (Christian Science, Yoga, Theosophy), and the serious interest taken by respectable scientists in psychic phenomena. Neo-Catholicism seems to have influenced Pierre Duhem's Thomistic philosophy of science. A version of irrationalism was developed by Bergson and Sorel in France, while German Idealism dominated the philosophy of Royce, Bradley, McTaggart, and the young Bertrand Russell.

Among all these diverse tendencies, we find a small group of theories that grew out of the dissatisfaction with classical physics and were to have an important influence on the philosophical interpretation of science in the twentieth century. They included positivism, empirio-criticism, energetics, and indeterminism, and they have some connection with the problems of the statistical interpretation of thermodynamics.

Positivism—this word has been used in various senses; it can denote simply the attitude that the methods of experimental science should be applied in sociology, history, and literature, instead of the methods of moral theology and metaphysics. In this sense it merges with "naturalism," "realism," and "materialism," and represents simply the influence of science on culture in the mid-nineteenth century. But in philosophy, and to some extent in science itself, positivism is opposed to materialism, and came to be linked with "idealism" or "empiricism." Positivists claim that materialism is a metaphysical doctrine which goes beyond the limits of observation by asserting that all phenomena derive from a material "substance," whereas positivism rejects all hypotheses about the unknowable (Littré 1864). Thus while positivists may advocate scientific methods in the study of society and the arts, they would place severe restrictions on the methods which can properly be used in science.

Positivism as a nineteenth-century movement derived from the writings of August Comte. While he believed in atoms (1830, vol. II, p. 206), he objected to the widespread use of supposed "agents" like the ether as a basis for classification of phenomena. His distinction between "fundamental" and "descriptive" sciences has been discussed in Chapter IV of this book. Comte became somewhat notorious among scientists for his dogmatic assertions about the impossibility of ever obtaining knowledge about the distances and chemical composition of the stars, made only a few years before such knowledge was in fact obtained through parallax measurements and spectroscopy. Thus positivism seemed to be a form of *negativism* in science!

Comte's philosophy was at first more influential in England (with the support of J. S. Mill) than in his own country, France. It was attacked by T. H. Huxley (1868), who said that he had found little of scientific value in positivism, "and a great deal which is as thoroughly antagonistic to the very essence of science as anything in ultramondaine Catholicism. In fact, M. Comte's philosophy, in practice, might be compendiously described as Catholicism *minus* Christianity" (Huxley 1871: 140, 152). According to Huxley's biographer Houston Peterson (1932), the phrase "Catholicism minus Christianity" had such a poisonous influence, regardless of Huxley's justification for it, that it was "one of the most damaging blows that the Positivists received in England" (Peterson 1932: 164), and helped to reduce positivism from a respectable intellectual position to the status of mere cultism.

Of the two senses of the word "positivism" mentioned above, the second one—a critical attitude toward scientific theories which attempt to go beyond the immediate facts of experience—became more common after 1870, and indeed positivism provided one component of the reaction against materialism with the object of reforming and reformulating the foundations of science. The older form of positivism came to be known as "scientism," an epithet connoting a misguided attempt to extend scientific terminology to subjects in which the scientific method cannot be applied. This pseudoscience merges with popular materialism, and many of the attacks on materialism are really attacks on scientism.

While scientific materialism would have set up boundaries within which the scientific method was applicable, but beyond which no sure knowledge was possible, the new positivism attempted to prove that the metaphysical mysteries of life and the universe supposedly lying beyond those boundaries are nonexistent. A distrust for metaphysical speculation was accompanied by suspicion of "intellectualism"; the scientific and mechanical models constructed by the human mind were denounced as unreal, frivolous, misleading, and worthless.

One casualty of this attitude was Gregor Mendel's atomistic theory of heredity. According to Bentley Glass (1953), a romantic distrust of any such mechanistic explanation of biological phenomena led scientists like Nägeli to reject Mendel's theory; it had to wait for the next realist period (see Chapter VIII) to find a sympathetic reception.

While the materialists had advanced physical explanations of biology, the positivists now emphasized instead biological explanations of physics. The latest researches in physiological psychology were adduced by Ernst Mach and others to prove that all perceptions, whether of color, smell, weight, or distance, are on an equal footing as combinations of sensations, and that there is no justification for choosing mass, distance, force, and time as physical quantities more fundamental than the others. If the human mind is, as the materialists asserted, merely another product of biological evolution, then obviously everything which has been developed by the human mind—including mechanical models and scientific materialism—is likewise simply a product of evolution.

Empirio-criticism is the doctrine of Avenarius, Mach, and their followers; it asserts that reality is a combination of sensations standing in a definite relation to each other. The true elements of the world are thus colors, sound, durations, pressures, smells, and spaces—not *things*. The idea of a thing-in-itself from which these qualities emanate is rejected as an illusory human thought-construct. A concept (or a name which appears to refer to a thing) is useful in summing up a set of experiences or sensations; and a scientific theory is likewise useful as long as it systematizes our concepts in a certain area. The goal of science is not to understand some mythical "reality" but to economize efforts of thought. Empirio-criticism held that the method of mechanical analogies had outlived its usefulness in science, because the attempts to explain phenomena by reducing them to hypothetical atomistic models had not led to simplicity but rather to greater complexity. In the case of heat, for example, the fact that heat can be transformed into mechanical work, and conversely, does not imply that heat has anything to do with mechanical processes or motion. (This was also the view of Robert Mayer.) It is quite irrelevant for scientific purposes whether we consider heat a substance or a mode of motion. That which we imagine to lie behind the appearances exists only in our minds and will vary from person to person, from culture to culture, and from time to time. (Poincaré's "conventionalism" has some affinities with this viewpoint.)

Ludwig Boltzmann, in his 1895 letter to *Nature* cited in Chapter VI, admitted that

> neither the Theory of Gases nor any other physical theory can be quite a congruent account of facts. . . . Certainly, therefore, Hertz is right when he says: "The rigour of science requires, that we distinguish well the undraped figure of nature itself from the gay-coloured vesture with which we clothe it at our pleasure." But I think the predilection for nudity would be carried too far if we were to forego every hypothesis (Boltzmann 1964:16).

The establishment of the conservation of energy was considered by many scientists to have put the kinetic theory on a legitimate foundation, and to suggest that ultimately all forms of energy might be reduced to the kinetic energy of matter in motion and the potential energy of forces between atoms. The school of *energetics* (Helm, Ostwald, Duhem, and others) maintained on the contrary that the equivalence of all forms of energy, far from authorizing us to reduce one of these forms to another, places them all on the same level. Whereas the empirio-critics would have accepted the reduction of all forms of energy to one form, provided that this reduction leads to a genuine economy of thought, the energetists were unwilling to accept any reduction at all. Instead, they assigned to energy itself the most fundamental reality. The appellation neoromantic is rather appropriate for energetics, since this doctrine is quite similar to the notion of unity of natural forces in *Naturphilosophie* which had preceded the establishment of the conservation of energy. Ostwald even founded a journal which he called *Annalen der Naturphilosophie*. We have already mentioned the argument of energetics against the mechanical theory, based on the second law of thermodynamics: energy is subject to irreversible dissipation in all natural transformations, a property which is not shared by a mechanical system of particles obeying Newton's laws of motion. Finding a contradiction between the absolute validity of thermodynamics and the kinetic theory, the energetists proposed to reject the latter.

Like many other aspects of neoromanticism, energetics was not especially popular in Britain. G. F. FitzGerald, reviewing Ostwald's work in 1896, said:

> The view of science which puts he forward—a sort of well-arranged catalogue of facts without any hypotheses—is worthy of a German who plods by habit and instinct. A Briton wants emotion—something to raise enthusiasm, something with a human interest. . . . This deadly science without hypothesis is far worse than the materialistic *ignorabimus* of Du Bois Reymond; it is the culmination of the pessimism of Schopenhauer (FitzGerald 1896:441).

It has recently been suggested, by followers of the "methodology of scientific research programmes" of Imre Lakatos, that Boltzmann's kinetic research program was degenerating in comparison to the thermodynamic approach in the 1880s and 1890s, and that even Boltzmann himself retreated from this program (Peter Clark 1976, Elkana 1974). But however one may interpret the philosophical pronouncements of Boltzmann and other scientists during this period, the fact remains that some of them did continue to do valuable research in the kinetic theory of gases and statistical mechanics (e.g. the work of Boltzmann and Jäger on the equation of state, Lorentz on the electron theory of metals, Gibbs on statistical mechanics and Smoluchowski on rarefied gases). According to Lakatos, the scientific community should not have tolerated research on such a "degenerating programme"—journals should have refused to publish these papers, and research foundations should have refused money (Howson 1976: 16). Fortunately the scientific community in the late nineteenth century was more tolerant of unfashionable theories than Lakatos thinks it should be; kinetic-theory research *was* published and was thus available to assist the revival of the subject after 1905.

In mathematics and mathematical physics, interest in form and structure began to replace atomistic analysis. The theory of groups made substantial advances in dealing with invariance and symmetry properties, including those of crystals. The dispute between Leopold Kronecker and Georg Cantor may be regarded as a conflict between neoromanticism and realism in mathematics. Cantor had proposed a theory of infinite sets of points which allowed one to conclude, for example, that the irrational numbers in a given line segment are more numerous than the rational numbers. But Kronecker insisted that "God Himself made the whole numbers—everything else is the work of men" and blocked Cantor's academic career. Even Poincaré thought Cantor's set theory was a disease from which mathematics must try to recover. After Kronecker died in 1891, Frobenius likened one of his early works to "chemistry without the atomic hypothesis" (Biermann 1973: 506).

Laymen were often alarmed by the deterministic attitude that scientists took toward the physical world; they supposed that science had adopted fatalism and rejected free will. Such statements as that of Laplace (1773, 1814), that the entire past and future history of the world could be predicted from its present state by a being with sufficient intelligence, were unacceptable to many. Some writers tried to reconcile free will with the principle of energy conservation (Croll 1872, 1891; Clark 1888; Lodge *et al.* 1891).

A few philosophers such as Cournot in France and C. S. Peirce in America postulated an element of chance or contingency in nature to limit the reign of determinism. The views of Peirce have some interest here because of their relation to the kinetic theory in both its classical and modern (i.e., quantum-mechanical) versions. The situation mentioned by Peirce is similar to that discussed by Zermelo and Boltzmann: if you have a theory that postulates that molecular motions are random, then how do you explain the fact that the behavior of macroscopic physical systems is so regular and deterministic? Why doesn't randomness show up on a large scale if it is really present on a small scale? The answer is that, because a huge number of molecules are involved, the probability of an observable spontaneous fluctuation from regular behavior is so small that it would not occur for many millennia. (At that time no one seemed to realize that Brownian movement depends on just this kind of spontaneous fluctuation.)

Peirce relied on the fact that the kinetic theory does postulate molecular randomness (as he wanted to do), yet one does not therefore expect to find any "tremendous effects" due to concentrations of heat. Hence there is no incompatibility between molecular randomness and macroscopic regularity. In 1892 he wrote (after noting the idea of Epicurus that atoms swerve randomly from their courses): "the peculiar function of the molecular hypothesis in physics is to open an entry for the calculus of probabilities" (1958: 162). Whereas determinists postulate that "certain continuous quantities have certain exact values" (1958: 169), anyone who has ever worked in a physical laboratory knows that it is impossible to determine any quantity by observation with zero error. One might think that if there were an element of randomness in the universe, it would occasionally produce effects that would be observed; yet the kinetic theory of gases is based on the assumption that atoms move about as if by chance, and

> by the principles of probabilities there must occasionally happen to be concentrations of heat in the gases contrary to the second law of thermodynamics, and these concentrations, occurring in explosive mixtures, must sometimes have tremendous effects. Here, then, is in substance the very situation supposed; yet no phenomena ever have resulted which we are forced to attribute to such chance concentrations of heat, or which anybody, wise or foolish, has ever dreamed of accounting for in that manner (Peirce 1958: 171).

Peirce thus supported Boltzmann's statistical interpretation of the second law of thermodynamics, though for not quite the same reasons. In another article, he doubted "whether the fundamental laws of mechanics hold good for single atoms," and suggested that instead of precise universal laws one should look for approximate laws of nature which may be the "results of evolution," thus introducing "an element of indeterminacy, spontaneity, or absolute chance in nature" (Peirce 1958: 148).

During the last half of the nineteenth century many criticisms of the kinetic theory appeared in the scientific and philosophical literature. Much of this criticism deserves no more than passing notice, since it came from writers who lacked the technical qualifications to make a sound judgment of the scientific value of the theory, and were simply reflecting the general reaction against materialism. I have already mentioned the criticisms based on the reversibility and recurrence paradox, which are of great scientific interest and also have philosophical implications.

The objections of J. B. Stallo may be viewed as a concrete application of the ideas of positivism and empirio-criticism mentioned above. His book *The Concepts and Theories of Modern Physics,* first published in 1882, is a remarkably penetrating analysis of the foundations of classical physics, which gained the admiration of Ernst Mach and went through fifteen printings in six different languages within thirty years of its original appearance. Stallo's systematic attack on the mechanical theory of the universe includes several substantial arguments against the validity of the atomic hypothesis and in particular the kinetic theory of gases, which may be summarized as follows:

1. If atoms are absolutely hard, they cannot be "elastic," since elasticity involves the motion of parts; in the collision of ordinary bodies there is a temporary loss of motion which is accounted for by conversion of energy of large-scale motion into energy of motion of the constituent parts, but this is impossible in the case of atomic collisions. The kinetic theorists, in order to maintain the principle of conservation of energy, had to assume that atoms are perfectly elastic, which is contradictory to the concept of an atom. (Wilson Scott [1970] gives a comprehensive history of this problem.)

2. The postulate that atoms are indestructible and inpenetrable cannot legitimately be inferred from experiences with ordinary solids, liquids, and gases.
3. A satisfactory scientific theory must explain obscure facts by means of familiar facts; "a valid hypothesis reduces the number of the uncomprehended elements of a phenomenon by at least one." (Quoted by Stallo 1960: 132–133; Zoellner 1872.) In the case of gas theory we have to explain first of all the existence of elasticity, i.e., resistance to compression, and the tendency toward expansion when external constraint is removed. Yet the kinetic theory proposes to explain this by invoking the supposed elasticity of invisible solid particles, which is more complicated and less comprehensible than the elasticity of gases, since a solid shows resistance to both compression and dilatation, and also to change of shape. In order to compensate for this defect and try to explain the tendency of a gas to expand, the kinetic theory must resort to still more hypotheses, which are likewise remote from experience. It is assumed that the atoms are endowed with incessant rectilinear motion, and that they exert no forces on each other except when in contact; but we have no knowledge of such behavior in the real world. Thus the kinetic theory only complicates the phenomena which it professes to explicate; it represents "an unraveling of the Simple into the Complex, an interpretation of the Known in terms of the Unknown, an elucidation of the Evident by the Mysterious, a reduction of an ostensible and real fact to a baseless and shadowy phantom" (Stallo 1960: 145).
4. The various artificial force laws introduced in order to account for certain properties of gases are "fatal to all claims of simplicity preferred on behalf of the kinetic hypothesis, and are in no sense an outgrowth of its original postulates. . . . They are both mere stop-gaps of the hypothesis, peace-offerings for its non-congruence with the facts, pure inventions to satisfy the emergencies created by the hypothesis itself" (1960:147–48).
5. There are no logical, mathematical, or other grounds for applying the statistical method to the *velocities* of the molecules instead of to their *weights* or *volumes*.

6. The theory fails to explain the relation between the thermal properties of gases and the internal motions of atoms in molecules.

In conclusion, Stallo says:

It may seem strange that so many of the leaders of scientific research, who have been trained in the severe schools of exact thought and rigorous analysis, should have wasted their efforts upon a theory so manifestly repugnant to all scientific sobriety—an hypothesis in which the very thing to be explained is but a small part of its explanatory assumptions. But even the intellects of men of science are haunted by prescientific survivals, not the least of which is the inveterate fancy that the mystery by which a fact is surrounded may be got rid of by minimizing the fact and banishing it to the regions of the Extra-sensible. The delusion that the elasticity of a solid atom is in less need of explanation than that of a bulky gaseous body is closely related to the conceit that the chasm between the world of matter and that of mind may be narrowed, if not bridged, by a rarefaction of matter, or by its resolution into "forces." The scientific literature of the day teems with theories in the nature of attempts to convert facts into ideas by a process of dwindling or subtilization. All such attempts are nugatory; the intangible specter proves more troublesome in the end than the tangible presence. Faith in spooks (with due reverence be it said for Maxwell's thermodynamical "demons" . . .) is unwisdom in physics no less than in pneumatology (Stallo 1960: 151).

A modern logical positivist could hardly have proved more conclusively the absurdity of belief in atoms.

Chapter VII

Degeneration

> A cloud was on the minds of men
>   and wailing went the weather
> Yea a sick cloud upon the soul
>   when we were boys together.
> Science announced nonentity
>   and Art admired decay
> The world was old and ended
>   but you and I were gay.
> —(Chesterton 1942: 109)

In 1857 there appeared in France two works whose effect on biology and literature, respectively, were comparable to that of the dissipation principle on geology. One was a scholarly medical treatise, Morel's *Traite des Dégénérescences Physiques, Intellectuelles et Morales de l'Espece Humaine*. The other was a collection of sensual poems, *Les Fleurs du Mal*, by Charles Baudelaire. The ideas of Morel and Baudelaire were destined to collide and fuse in one of the most bizarre doctrines of the neoromantic period: the theory of degeneration.

Morel, two years before the publication of Darwin's *Origin of Species*, proposed a theory of retrograde evolution which furnished a dismal counter-current to the predominantly optimistic outlook of Darwinism. Morel defined "degeneration" as a morbid deviation of an organism from its original type. His theory was further developed by Lombroso and others, and attained considerable popularity in certain circles during the last part of the nineteenth century. It appeared to provide a comprehensive scientific explanation for all the evils of society by attributing them to individuals afflicted from birth by idiocy, criminality, moral depravity, insanity, physical deformity, and other symptoms of degeneration. It was assumed that these characteristics were inherited from parents who possessed them to a much smaller extent, but whose constitutions had been weakened by indulgence in alcohol and drugs, overwork, and similar debilitating influences. This is not quite a theory of "inheritance of acquired characteristics" (the heresy which Darwinists are always fighting) but it does postulate that the environment of one generation can affect the hereditary endowment of the next. The supposed increase in degeneration in the human race is explained as a result of the increasingly frantic conditions of modern life, which produce nervous exhaustion and fatigue, together with a craving for stimulants.

The detailed etiology of degeneration varied from one country to another. In England, deterioration was due to "the lowering proportion of the Nordic blood and the transfer of political power from the vigorous aristocracy and middle classes to the radical and labor elements, both largely recruited from the Mediterranean type" (Grant 1917: 186). W. K. Clifford (1876) pointed to the degenerating effects of Christianity as a social system in Malaga. In the United States, where observers such as George M. Beard had called attention to a peculiar "American nervousness" as early as 1881, conditions were thought to be favorable for the production of degenerates. G. F. Lydston (1904: 36) suggested that "Lust for wealth, desire for social supremacy, ambition for fame, love of display, late hours, lack of rest, excitement—all these factors combine to cause . . . a distinctively American disease. The body social is growing more and more neuropathic. In the train of this widespread neuropathy comes degeneracy, with all its evil brood of social disorders." The related affliction of "neurasthenia" was most likely to strike the liberated "new woman" according to many physicians (Haller 1971). Fear of racial degeneration was also one reason for the white Americans' dread of miscegenation.

But it was especially in France that degeneration was observed on all sides; the population was declining, foreigners were becoming more numerous, crime was increasing, families were becoming disorganized, and the nation was obviously losing its political leadership in Europe.

Degeneration found a permanent place in literature through the novels of Émile Zola. Zola was a realist rather than a neoromantic; his writings were directly influenced by science, in particular by an early work of Prosper Lucas on heredity (1847). In his notes for the Rougon-Macquart novels, Zola mentioned:

> Exhaustion of the intelligence, through the rapidity of the flight toward the heights of sensation and thought. Return to demoralization. Influence of the feverish modern environment on the ambitions of impatient individuals. . . . I am studying the ambitions and appetites of a family launched into the modern world, making superhuman efforts and not succeeding, because of its own nature and certain influences; touching success only to fall back; and ending by producing veritable moral monstrosities (the priest, the murderer, the artist). The time is troubled; it is the trouble of the time that I am painting (Josephson 1928: 145).

Elsewhere Zola said even more explicitly: "Tous trois, Serge, Désirée, Octave, sont d'ailleurs des dégénérescences. Il faut les étudier à ce point de vue" (Martineau 1907: 110). In the interests of science Zola even permitted himself to be examined by a specialist in degeneration, who reported

> Zola is neither epileptic nor hysterical, nor is there the least sign of mental alienation. Although he has many nervous troubles, the term "degeneracy" does not apply to him wholly. Magnan classes him among those degenerates who, though possessing brilliant faculties, have more or less mental defects . . . (MacDonald 1898: 494).

After France was defeated by Germany in the war of 1870, it was feared that the Latin races had become decadent and would henceforth be dominated by the Nordic races. Around 1880 a circle of French poets began to call themselves the Décadents. The name was apparently derived from Théophile Gautier's 1868 preface for Baudelaire's *Fleurs du Mal;* the Decadents were, among other things, the imitators of Baudelaire.

Baudelaire's own feeling for degeneration is exemplified by the following excerpt from his poem "Une Charogne":

>   Do you remember the thing we saw, my soul,
>   That summer morning, so beautiful, so soft:
>   At a turning in the path, a filthy carrion,
>   On a bed sown with stones,
>
>   Legs in the air, like a lascivious woman,
>   Burning and sweating poisons,
>   Opened carelessly, cynically,
>   Its great fetid belly.
>
>   The sun shone on this fester,
>   As though to cook it to a turn,
>   And to return in a hundred pieces to great Nature
>   What she had joined in one . . .
>
>   * * *
>
>   —And yet you will be this excrement,
>   This horrible stench,
>   O star of my eyes, sun of my being,
>   You, my angel, my passion.
>
>   Yes, such you will be, queen of gracefulness,
>   After the last sacraments,
>   When you go beneath the grasses and fat flowers,
>   Mouldering amongst the bones.
>   —(Baudelaire 1946: 49)

The Aesthetes in England shared many of the viewpoints of the Décadents and Symbolists in France. Oscar Wilde's *The Picture of Dorian Gray* was attacked by a reviewer who called it a "tale spawned from the leprous literature of the French *décadents*—a poisonous book the atmosphere of which is heavy with the mephitic odours of moral and spiritual putrefaction." Wilde replied: "It is poisonous, if you like, but you cannot deny that it is also perfect, and perfection is what we artists aim at" (quoted by Mason 1914: 48).

In Germany, Richard Strauss set Wilde's play *Salome* to music, thus producing what one historian has called "a music-drama which has probably no parallel in its excitement of disgust and sheer horror. . . ." The heroine "glows with a phosphorescent light that almost makes us mistake putrefaction for health" (Ferguson 1935: 469). Strauss and Debussy undermined tonality, thus preparing the way for the collapse of romantic music at the beginning of the twentieth century. Mahler's *Das Lied von der Erde* (1906) has been called the "swan song of the entire Romantic movement" (Copland 1968: 32).

It is not surprising that some writers suggested a relation between biological degeneration and artistic decadence or neoromanticism. It was Max Nordau, a Hungarian physician and social critic, and later a Zionist, who carried the comparison to an extreme by proposing that many of the leading artists and writers of the day were themselves actually degenerate individuals, and that their work and personal lives were symptons of degeneracy. In his book *Entartung* (1892) he develops this theory at length, and provides us with a fascinating description of the neoromantic movement in the arts. He begins by quoting various uses of the phrase *fin-de-siècle* in France, concluding that the common feature of all the usages is "a contempt for traditional views of custom and morality" (Nordau 1968: 5). To elaborate:

> It means a practical emancipation from traditional discipline, which theoretically is still in force. To the voluptuary this means unbridled lewdness, the unchaining of the beast in man; to the withered heart of the egotist, disdain of all consideration for his fellow-men, the trampling under foot of all barriers which enclose brutal greed of lucre and lust of pleasure; . . . to the believer it means the repudiation of dogma, the negation of a supersensuous world, the descent into flat phenomenalism; to the sensitive nature yearning for aesthetic thrills, it means the vanishing of ideals in art, and no more power in its accepted forms to arouse emotions. . . .
>
> One epoch of history is unmistakably in its decline, and another is announcing its approach. . . . Things as they are totter and plunge, and they are suffered to reel and fall, because man is weary, and there is no faith that it is worth an

effort to uphold them. . . . Over the earth the shadows creep with deepening gloom, wrapping all objects in a mysterious dimness, in which all certainty is destroyed and any guess seems plausible. Forms lose their outlines, and are dissolved in floating mist. The day is over, the night draws on . . . (Nordau 1968: 5–6).

The symptoms of the *fin-de-siècle* state of mind, according to Nordau, are exhibitionism in dress, proliferation of strange colors in painting, dissonance or fake religiosity in music, obscurity and mysticism in literature. These are evident to everyone, but "the purely literary mind, whose merely aesthetic culture does not enable him to understand the connections of things, and to seize their real meaning," merely talks loftily of a "restless quest of a new ideal by the modern spirit" (Nordau 1968: 15). But the physician recognizes in the *fin-de-siècle* disposition, and in the tendencies of neoromantic art and poetry, "the confluence of two well-defined conditions of disease, with which he is quite familiar, viz. degeneration (degeneracy) and hysteria" (*ibid.*). If one could submit the originators of the neoromantic movement to a careful physical examination, together with their relatives, one could probably discover one or more stigmata which would confirm the diagnosis of degeneration; but this humiliating procedure is not necessary, for there are other symptoms of degeneration which can be found in the artistic productions themselves, as well as in the public appearance and behavior of the artists. Among these are: moral insanity, emotionalism, mental weakness, and pessimism; an unconscious fear of everything and everyone; a disinclination to action of any kind; a predilection for inane reverie. The degenerate is often "tormented by doubts, seeks for the basis of all phenomena, especially those whose first causes are completely inaccessible to us"; he supplies "new recruits to the army of system-inventing metaphysicians, profound expositors of the riddle of the universe, seekers for the philosopher's stone, the squaring of the circle and perpetual motion" (1968: 21). He may be an anarchist, or he may devise "plans for making mankind happy" (1968: 22). A cardinal mark of degeneration is mysticism. But it must not be thought that the degenerate individual is necessarily stupid; on the contrary, he may be a genius, or have great artistic talent, although in such cases he usually has one gift exceptionally developed at the cost of the remaining faculties.

While it is the degenerates who originate art, it is the hysterics who eagerly acclaim and popularize it. Together they tend to form close groups or schools, uncompromisingly exclusive to outsiders. This by itself is an unnatural feature of modern art, according to Nordau, since "If any human activity is individualistic, it is that of the artist. True talent . . . follows its creative impulses, not a theoretical formula preached by the founder of a new artistic or literary church" (1968: 29). But "hypersusceptibility to suggestion is the distinguishing characteristic of hysteria. . . . When a hysterical person is loudly and unceasingly assured that a work is beautiful, deep, pregnant with the future, he believes in it" (1968: 32). Both the degenerates and their hysterical followers are "sincere" insofar as they "act as, in consequence of the diseased constitution of their brain and nervous system, they are compelled to act" (1968: 31). However, the movement may succeed in also attracting to itself unbelievers who hope to acquire fame and money as associates of the new sect even though they recognize its insanity.

Nordau counts realists and naturalists as degenerates, in addition to decadents and symbolists. Romanticism itself, he asserts, is a manifestation of degeneration, though he regards Goethe as a well-balanced healthy artist. He sees symptoms of degeneracy in the pre-Raphaelites, the Oxford Movement, Ruskin, neo-Catholicism, Baudelaire, Verlaine, Tolstoi, Wagner, the Rosicrucians, Maeterlinck, Swinburne, Oscar Wilde, Ibsen, Nietzsche, Zola, and many others. Yet he is optimistic, for according to theory, the degenerate lines must succumb because they cannot adapt themselves to the conditions of nature and civilization, while the rest of the race will adapt itself and become stronger.

> The end of the twentieth century, therefore, will probably see a generation to whom it will not be injurious to read a dozen square yards of newspapers daily, to be constantly called to the telephone, to be thinking simultaneously of the five continents of the world, to live half their time in a railway carriage or in a flying machine, and to satisfy the demands of a circle

of ten thousand acquaintances, associates, and friends. It will know how to find its ease in the midst of a city inhabited by millions, and will be able, with nerves of gigantic vigour, to respond without haste or agitation to the almost innumerable claims of existence (Nordau 1968: 541).

Of the various responses to Nordau's onslaught, only George Bernard Shaw's "The Sanity of Art" seems worth mentioning here; anyone who takes the time to look up *Degeneration* should not fail to read Shaw's pamphlet.

Though it has now been forgotten by the public and is ignored in "Whig" histories of science, the theory of degeneration did play an important role in twentieth-century social history because of its influence on two "progressive" reform movements: eugenics and prohibition. Through eugenics, it was also involved in the arguments for immigration restriction in the 1920s. Allen (1976) gives a useful review of recent historical writing on these movements.

The eugenics movement was initiated in England by Francis Galton (1822–1911), though some of its ideas go back to Plato. After publishing influential works on heredity such as *Hereditary Genius* (1869) and experimenting with various tests of mental ability, Galton founded the Eugenics Education Society in 1908; this Society published the *Eugenics Review*, which was widely read. In his will, Galton provided £45,000 for a eugenics laboratory at the University of London, and his student Karl Pearson carried on important work, oriented especially toward the use of statistical methods. The journal *Biometrika* was founded by Galton and Pearson in 1901.

Galton believed that intelligence was primarily determined by heredity, though he did not exclude the possibility of environmental effects on heredity. He thought that the superior people in a society had a responsibility to marry other superior people and reproduce their kind. Contrary to Malthus, Galton did not advocate voluntary birth control; on the contrary he pointed out that birth control was more likely to be practiced voluntarily by superior people and advanced races, who would thereby be committing race suicide since the lower classes and inferior races would continue to propagate without restraint. He also opposed the Roman Catholic concept of chastity, which decreased the level of intelligence by preventing the more intelligent people (who became priests and nuns) from reproducing.

Galton and his followers attempted to identify various factors in the environment or in society that could be considered "racial poisons" tending to favor the reproduction of stupidity rather than intelligence. For example, the movement of population from the rural areas into the cities was considered to have bad effects. Galton pointed out that north British peasant women looked happy while the urban masses, especially women, seemed miserable. The cities tended to attract the brightest and most energetic people, yet once there, they and their descendants tended to deteriorate; hence urbanization was a "racial poison" that degraded the entire race. This contention harmonized with numerous writings of the late nineteenth century decrying the effects of city life on the health and mental stability of people, and promoting nostalgic idealizations of country life.

By eliminating racial poisons and encouraging the breeding of better humans, the eugenists proposed to control the evolution of the human race. This program turned out to be very attractive in the United States, where it was incorporated into the thinking of many of the Progressives in the early decades of the twentieth century. It also provided "scientific" support for the prejudices of people who were interested in passing various kinds of restrictive legislation (e.g., Prohibition and Immigration laws) even though they did not accept the theory of evolution.

The leader of the American eugenics movement was Charles Davenport (1866–1944), a zoologist who did some solid research work in embryology and genetics, and pioneered the use of Karl Pearson's statistical ("biometric") methods in the United States. In 1904 Davenport persuaded the Carnegie Institution to give financial support to the Station for Experimental Biology at Cold Spring Harbor, New York, and became its director. He became interested in eugenics at the instigation of his wife, Gertrude Crotty Davenport, who collaborated on several papers with him. He collected data on the inheritance of various traits in families. With further support from Mrs. E. H. Harriman (who reportedly thought that experience in breeding race horses should be applied to humans), he set up the Eugenics Record Office, which later became part of the Carnegie Institution (in 1934 its name was changed to the Genetics Record Office).

Davenport proposed some rather mild forms of eugenic policies: people should be careful in selecting their marriage partners; superior couples should have larger families; the races should not be intermixed; and "undesirable" immigrants should be excluded from the United States.

Another leader of the eugenics movement was Paul Popenoe, who wrote popular books on eugenics and marriage. It is interesting to note that Popenoe could be considered a typical member of the "Progressive" movement despite his advocacy of many positions that seem quite conservative today: he considered socialism a menace because it ignored the basic and natural inequality of mankind; he questioned the wisdom of child-labor laws, minimum wage legislation, mothers' and old-age pensions, and trade unionism, all of which he claimed contributed to preserving the biologically and mentally inefficient members of the race. He defended antimiscegenation laws and insisted on the inferiority of Negroes. He encouraged the back-to-the-farm movement, arguing that the city undermined morality and allowed the increase of the inefficient; and he felt that experts rather than democratically elected officials should determine public policy. In all this it is difficult to disentangle the scientific justifications from the conservative bias.

The American eugenics movement eventually became best known for its advocacy of sterilization; and this led to its downfall. The idea that people afflicted by what seemed to be hereditary defects—feeblemindedness, incurable diseases, criminality, etc.—should not be allowed to have children had been proposed several times in earlier centuries but had never gained approval as a public policy. In the 1890s, surgical sterilization methods (e.g., vasectomy in men, salpingectomy in women) had been developed. Unlike castration, these methods appeared to be safe and simple and did not interfere with normal sexual functioning or the development of the secondary sexual characters. The eugenists demanded that defective persons be sterilized in order to halt the deterioration of the race. In 1905 a sterilization bill was passed by the Pennsylvania legislature but was vetoed by the governor; in 1907 a law was put into effect in Indiana, requiring sterilization of inmates of state institutions who were insane, idiotic, imbecilic, feebleminded, or who were convicted rapists or criminals, whenever recommended by a board of experts. By 1931, thirty states had passed similar laws.

Although there was some criticism of the eugenists by scientists who felt that the state of knowledge of human genetics did not yet justify taking such drastic measures, the biggest blow to the eugenics movement came not from its enemies but from its friends. Adolf Hitler was an advocate of eugenics, and when the Nazis came to power in Germany in the 1930s they quickly passed a law providing for sterilization to carry out their program of "race hygiene." During the first year of operation of this law, 56,244 persons judged to be "hereditarily defective" by the eugenic courts were sterilized. This was of course the first step toward the proposed "final solution" of the Jewish problem—the perpetuation of Aryan superiority by elimination (by sterilization or murder) of non-Nordic parts of the population.

By this time the American eugenists had already lost the support of most geneticists because they had failed to take account of current scientific research and seemed to have devoted themselves to racist propaganda. Popenoe was so enthusiastic about sterilization that he initially supported the Nazi program, not realizing what it would lead to. So "eugenics" became a dirty word, and even today any proposals for sterilization, voluntary or otherwise, are likely to be condemned as "genocide."

The collapse of the eugenics movement in the late 1930s had two unfortunate effects. First, the public came to be highly suspicious of even those aspects of the program that might have been of some value, if based more firmly on modern theories of heredity. Second, many scientists chose not to do research in *human* genetics because that field had acquired a bad reputation by its association with eugenics.

As mentioned before, many of the eugenists wanted to prevent race mixing by restricting immigration into the United States of persons of "inferior stock." There had been a huge wave of immigration from Ireland, Italy, eastern Europe, and other countries in the nineteenth century, and the "WASP" upper middle class felt itself threatened by the political power being acquired by the big-city "bosses" relying on the "ethnic vote." The eugenic arguments probably didn't convert anyone who was previously in favor of immigration, but they did give an aura of scientific authority to those who were urging restrictions.

With the advice of several members of the eugenics movement, Albert Johnson, Chairman of the House Committee on Immigration and Naturalization, proposed a bill to limit the annual number of immigrants from each European country to not more than 2% of the United States residents listed in the 1890 census who had been born in that country. This became the Immigration Restriction Act of 1924, and was in effect until it was replaced by the Celler Act of 1965 which removed the discrimination against particular countries. The net effect of the Johnson Act, as intended, was to reduce the immigration from southern and eastern Europe, on the theory that inferior races inhabited those countries. It also discriminated against the Irish who, though from northwest Europe, were considered inferior because they were Catholic.

Another "social reform" associated with eugenics was the "temperance" movement leading to legislations prohibiting the sale of alcoholic beverages on the state and national level. Historians who have discussed the causes of Prohibition—i.e., the reason why it was actually put into effect in the United States in 1919—have not found it difficult to identify a number of factors present in the early twentieth century: the Progressive movement, World War I, the corruption of the liquor interests, etc. I will not discuss the arguments about those factors but consider only the way in which the results of scientific research, as presented to the public, might have influenced the success of the temperance movement at this particular time.

In looking for "scientific" justifications of Prohibition we must recognize from the outset that much of the support for it came from the South and from segments of the population where fundamentalist religion was strong, and where the prestige of science was weak or nonexistent. Thus William Jennings Bryan, who was one of the leading politicians involved in the movement, would not have considered scientific theories as having any value in the debate, especially if they had any connection with a theory of evolution. (He was a principal figure in the notorious 1925 "monkey trial" in Tennessee.) Just as the true racist was often reluctant to invoke evolutionary arguments about the biological status of blacks if they might conflict with his religious beliefs, so the diehard prohibitionist should have been careful about the implications of his more extreme claims concerning the possible effects of alcohol on humans. Nevertheless some prohibitionists did actually invoke evolutionary arguments, except that they referred to "negative" or "backwards" evolution: the theory of degeneration.

In interpreting the effects of scientific ideas on political behavior such as that involved in the enactment of Prohibition, we have to deal with the "swing vote"—the voters, legislators, and opinion-makers who (as a group) changed their views from anti-Prohibition before 1915 to pro-Prohibition between 1915–25, and back to anti-Prohibition after 1925. I will not try to identify the members of this group, except to note that they must have existed, since the national policy on Prohibition *did* change rather dramatically two times in two decades. I am not concerned with the hard core who remained firmly committed to one side or the other all the way through, but rather with those people who changed their minds. My hypothesis is that these people were the ones who would ordinarily be opposed to any state interference with an individual's right to damage himself by behavior such as drinking, unless there were good reason to suppose that he was thereby damaging society as well. With this in mind, let's review what scientists and physicians said about the effects of alcohol in the late nineteenth and early twentieth centuries.

Before 1860 there was little scientific evidence on the physiological effects of alcohol; popular notions of the *beneficial* effects of liquor were reinforced by the practice of physicians who often prescribed it for heart trouble and other diseases. However, medical research late in the nineteenth century revealed that alcohol is a depressant rather than a stimulant, and may have many harmful effects on various organs. One finds an enormous amount of medical literature on alcoholism from countries such as Sweden, Germany, France, and England. Perhaps this was an effect of urbanization and the industrial revolution, which made alcohol more freely available to many people and removed the previous taboos against overindulgence.

As it happened, the degeneration theory of Morel and his followers developed in close association with clinical and causal studies of alcoholism. Alcohol was very frequently cited as a primary cause of hereditary degeneration. A typical statement one finds is that out of 350 idiots examined in an insane asylum, 99 had drunken fathers (Howe 1858). The conclusion is that the parent's consumption of alcohol affects the germ plasm in such a way as to cause the offspring to be defective. In addition to criminality and idiocy, the offspring will also be liable to become alcoholic himself, so that the effect is accentuated in the next generation.

Americans were especially susceptible to the argument that alcohol causes degeneration because of their experience with the Indians, on whom it produced such startling effects. Then, after 1865, numerous reports of the degeneration of Negroes freed from slavery emphasized the role of alcohol. This native experience was reinforced with all the prestige of European science in the 1890s and 1900s, and the temperance movement was able to give numerous learned citations from German and Swedish physicians in its propaganda.

There were some counter-arguments. One was that by weeding out the unfit, alcohol is actually good for the race in the long run; it is to be considered part of the "environment" that selects out the weaker members for extinction, leaving behind only those who have inherited a strong enough constitution to withstand its effects. From another angle, the neo-Darwinians following Weismann claimed that "alcoholic degeneration" is in principle impossible, since nothing that happens to an individual can possibly affect the germ plasm that determines the hereditary endowment of his offspring. The neo-Darwinians would claim that all the evil effects on children of alcoholic parents are purely environmental—that a child raised in such a home is naturally prone to all kinds of nervous disorders.

One of the first fruits of the new biometrical methods was the study published in 1910 by Ethel Elderton and Karl Pearson at the Galton Eugenics Laboratory in London. They studied "the influence of parental alcoholism on the physique and ability of the offspring" by examining schoolchildren in Edinburgh and Manchester, and sending social workers to investigate the family circumstances. They reported that, contrary to the expectations of the social workers who collected the data, "no *marked* relation has been found between the intelligence, physique or disease of the offspring and parental alcoholism. . . . On the whole the balance turns as often in favour of the alcoholic as of the non-alcoholic parentage. It is needless to say that we do not attribute this to the alcohol but to certain physical and possibly mental characters which appear to be associated with the tendency to alcohol" (Elderton and Pearson 1910: 31). Thus for example alcoholic parents tend to be more fertile; that is not *because* they drink but because fertility happens to be correlated with other factors that may favor alcoholism. They also found that children of drinking parents have slightly better eyesight than those of sober parents, which they attributed to the tendency of alcoholic parents to drive their children into the streets; the outdoor environment is better for the eyes. They suggested that there may indeed be a hereditary connection between alcoholism in parents and in children in that a *tendency to alcoholism* may be present in the germ plasm of both—but the child will inherit this tendency whether his parents actually drink or not. So they concluded that eugenists should focus on eliminating the defective members of society who have tendencies toward alcoholism—"cut off its source at the production of degenerative stocks" rather than simply prohibit the use of alcohol by everyone (Pearson and Elderton 1910: 1–2).

From the modern viewpoint the Elderton-Pearson study was vastly more sophisticated and reliable, as scientific research, than the earlier studies quoted as arguments for Prohibition. As Elderton and Pearson themselves pointed out, the other studies involved the common fallacy of confusing statistical correlation with a cause-effect relation, and ignoring other reasons that could account equally well for the correlations. Thus rather than assume that alcoholism causes degeneration, one must consider the (theoretically more plausible) possibility that alcoholism is a symptom of preexisting degeneration. But the Prohibitionists were unmoved by all this, if in fact they were aware of Elderton and Pearson at all. Popenoe and Johnson's *Applied Eugenics* (1918) did contain a careful review of the literature, and concluded that Elderton & Pearson's result was probably valid; at least that eugenics "must proceed on the basis that there is no proof that alcohol as ordinarily consumed will injure the human germ-plasm[;] . . . alcoholism may be a symptom, rather than a cause, of degeneracy" (1918:60–61). In fact, if there were really any long-term cumulative effects of alcoholism, of the type alleged by the Prohibitionists, the race would be already extinct!

But by 1918 it was too late to stop Prohibition. The 18th Amendment had been passed by Congress and submitted to the states for ratification on December 18, 1917, and on January 16, 1919, the 36th state (Nebraska) ratified it so that it went into effect a year later. The Volstead Act (providing for enforcement of the Amendment) was passed by Congress in October 1919, vetoed by President Wilson, passed again over his veto, and went into effect on January 17, 1920.

As evidence for the role of the "degeneration" argument in the success of the Prohibition movement, I quote from an influential speech by Richmond Pearson Hobson of Alabama, delivered in 1916 and reprinted in *Congressional Record*—Hobson was one of the leaders of the movement.

> My attitude on this great question has always been and it remains purely scientific. . . . The conclusions of modern science about alcohol . . . are final, just as firmly established as the law of gravity or the shape of the earth . . . (Hobson 1917: 7822).

There is then a review of two major findings of science about the direct effect of alcohol on the human body.

> . . . it remained for finding No. 3 to startle the world and cause the civilized nations to tremble with the realization that they must become sober or perish. The finding that alcohol causes degeneracy in all living things is in reality the specific for degeneracy, the process that entails extinction. . . .
>
> Modern science research has made the startling discovery that whether in the vegetable kingdom, the kingdom of the

animals, the kingdom of men, when ethyl alcohol is applied internally, whether to the single elemental life cell or to the complete organism, the evolutionary building process is arrested and finally reversed. This backward course of degeneracy, reversing the process of evolution, is the deadly sin in the eyes of nature. When degeneracy sets in, it is certain that a deadly blow has been struck at the fundamental processes of life ... so the life itself will be shortened and the offspring blighted in proportion to the degree of degeneracy. . . .

If both parents are alcoholic, one child in five becomes insane, one child in seven is born deformed, one child in three is backward, with a likelihood of becoming epileptic or feeble-minded; one child in six will be normal, and even this one may transmit a taint to its offspring.

The great issue of our age—the greatest issue of all ages—is to cut the millstone of degeneracy from the neck of humanity. Upon this issue will hinge the destiny of the race . . . (Hobson 1917: 7825).

It has sometimes been said that women were a major force supporting the Prohibition movement even though they did not yet have the vote. As an extreme example of the appeal based on degeneration, here is Carry Nation in her autobiography (1908):

The curse of heredity is one of the most heart-breaking results of the saloon. Poor little children are brought into the world with the curse of drink and disease entailed upon them. . . . If girls were taught that a drunkard's curse will in the nature of things include his children and also that if either parents allowed bad thoughts or actions to come into their lives, that their offspring will be a reproduction of their own sins, they would avoid these men, and men will give up their vice before they will give up women (Nation 1908: 75).

The theory of degeneration faded away in the early decades of the twentieth century, because of the ascendancy of theories that drew a sharper distinction between the effects of heredity and environment. August Weismann combatted Lamarckism in the 1880s and 1890s with his experiments designed to prove that "acquired characters" cannot be inherited, and the revival of Mendelian genetics in 1900 seemed to confirm his doctrine that the hereditary germ plasm is well insulated from environmental dangers such as alcoholism and immoral behavior of the parents. The statistical studies of Elderton and Pearson mentioned above were eventually accepted as proof that the decision to drink or not to drink has no direct influence on the hereditary endowment of one's offspring, even though it may be difficult to disentangle such effects from those of the family environment created by a drunkard, or genetic factors that predispose people toward both alcoholism and physical deterioration or mental retardation. Indeed, this was the first time that the sophisticated techniques of modern statistical analysis, developed by Karl Pearson and others around 1900, were applied to a controversial social issue.

The psychoanalytic doctrine of Sigmund Freud may be considered the successor to the theory of degeneration as an explanation of the cause of neuroses and deviant behavior. In his early writings, Freud criticized the psychiatric theories of Jean Martin Charcot and others, based on the concept of degeneration, as being a way of stigmatizing rather than explaining neuroses. Instead, Freud argued that while the hereditary endowment of the individual may be a precondition for mental life, it cannot be taken as a specific cause for the actual form of disorders; those specific causes must instead be sought in the sexual experiences and fantasies of the young child. Freud's theories may thus be regarded as the first stage in a general shift away from hereditary toward environmental explanations in the behavioral and social sciences.

Chapter VIII

The End

> The new artists have been violently attacked for their preoccupation with geometry. . . . But it may be said that geometry is to the plastic arts what grammar is to the art of the writer. Today, scientists no longer limit themselves to the three dimensions of Euclid. The painters have been led quite naturally, one might say by intuition, to preoccupy themselves with the new possibilities of spatial measurement which, in the language of the modern studios, are designated by the term: the fourth dimension.
> —(Apollinaire 1913, in 1949: 13)

The first and perhaps the only major writer to gather together all the strands of thought mentioned in this essay was the American historian Henry Adams. Adams had criticized Lyell's uniformitarian geology in the *North American Review* in 1868, and subsequently learned of Lord Kelvin's thermal geophysics from his friend Clarence King. He advised his brother Brooks Adams while the latter was writing his book *The Law of Civilization and Decay*, which detailed the economic decadence of modern Western civilization. Brooks Adams, a follower of Galton and Lamarck, asserted that:

> the peculiarities of mind are apparently strongly hereditary, . . . as the external world changes, those who receive this heritage must rise or fall in the social scale, according as their nervous system is well or ill adapted to the conditions to which they are born. Nothing is commoner, for example, than to find families who have been famous in one century sinking

into obscurity in the next, not because the children have degenerated, but because a certain field of activity which afforded the ancestor full scope, has been closed against his offspring... (Adams 1896: vii).

The parallel with the declining fortunes of the Adams family itself was apparent, though mentioned only in private: "It is now full four generations since John Adams wrote the constitution of Massachusetts. It is time that we perish. The world is tired of us" (Samuels 1964: 130).

In 1890 Adams went to the South Seas and observed the contrast between the healthy nudity of Samoa and the Westernized degeneration of Tahiti. Together with his brother Charles Adams, he studied signs of decay in his own father. He read Nordau's *Degeneration*, and Clarence King suggested that they "go and pose for Nordau together—he seems to have had no degenerates or hysterics of our type—fellows who know all about it but manage to get a world of fun and some pleasure from it" (Samuels 1964: 167). Visiting Rodin in Paris in 1895, he hesitated to buy one of the artist's bronze figures because "they are mostly so sensually suggestive that I shall have to lock them up when any girls are about, which is awkward; but Rodin is the only degenerated artist I know of, whose work is original" (1964: 415).

The eternal return had little meaning for Henry Adams, aside from his infatuation with the Middle Ages. There is no indication that he followed his brother Brooks, and the British diplomat Cecil Spring Rice, in urging Theodore Roosevelt to emulate Kaiser Wilhelm's new militaristic Germany, which, like a nation of Nietzschean Supermen, was preparing to purge the decadent Western world. In his own personal reaction against materialism, Adams wished to go backwards rather than forwards, and considered optimism a sign of idiocy.

Adams, like many modern intellectuals, both loved and hated science. He read science furiously, looking for weak points in the mechanistic system, and for analogies on which he could base his own theory of history. He twisted to his own purposes Maxwell's demon, the kinetic theory of gases, the laws of thermodynamics, Ostwald's energetics, and the dynamo; and took comfort in Lord

Kelvin's admission that he knew no more of the underlying behavior of matter than he had fifty years before. He was no more satisfied with the scientific critics of mechanism than with the supporters of it; of Stallo, he said: "Singular that the result of eliminating metaphysics should always be to become more metaphysical" (Samuels 1964: 385).

In the 1890s Adams began collecting evidence for all kinds of dissipation—physical, biological, geological, and social. In three essays, later published under the title *The Degradation of the Democratic Dogma,* he argued that a science of history could be based on the general properties of energy and entropy as discovered by the physical sciences. A useful introduction to this book by his brother Brooks Adams describes the disillusion with American democracy which forms the historical background for Henry Adams' ideas.

In the first essay (1894), Adams noted the recent change from the cheerful optimism inspired by Darwin to a general feeling of pessimism, and warned his colleagues that they would have to deal with the dismal future as well as the past in their teaching of history.

The second essay, written in 1909, was an attempt to apply the "Rule of Phase" of Willard Gibbs (1876) to history. Adams' reasoning is rather confused and confusing, but the basic idea can be explained quite simply. Equilibrium between two phases of the same substance—for example, ice and water—is possible only at a single temperature, if the pressure is fixed; the transition from one phase to the next is *discontinuous* with respect to temperature. As ice is heated, its temperature rises until the melting point is reached; then, a large amount of heat must be added to change all the ice to water, but the temperature will not begin to rise again until the transition has been completed. The energy of the system, which has been increasing at a fairly steady rate as the temperature goes up, suddenly jumps up to a new level at the temperature of the phase transition.

In addition to melting, there is another phase transition at a higher temperature, in which liquid changes to gas. Adams quoted the conjectures of physicists to the effect that there may be still further such phase transitions at higher temperatures, the other phases being electricity, aether, space, and hyper-space.

According to Adams, history may be interpreted by regarding thought (or human society in general) as undergoing successive phase transitions as time goes on. The rule of phase means nothing more than that such changes of phase are discontinuous; two different phases cannot coexist over any significant length of time. In history, such changes of phase are to be recognized by means of the marked change in direction and form of thought which accompany them (for example the Renaissance).

In order to construct a quantitative time scale for this succession of historical phases, Adams notes that the rate of change has been increasing during the last 300 years, and "the acceleration suggests at once the old, familiar law of squares" (Adams 1958: 285). Adams thus arrived at the following scheme: the Mechanical Phase is taken to have lasted 300 years (1600–1900), and it was preceded by the Religious Phase, whose beginning cannot be determined but, for the sake of illustrating the hypothesis, may be assigned a length of $300^2 = 90{,}000$ years. The next or Electric Phase would last $\sqrt{300} \approx 17\ 1/2$ years; it would then, in 1917, pass into the Ethereal Phase, "which, for half a century, science has been promising, and which would last only $\sqrt{17.5}$ or about four years. and bring Thought to the limit of its possibilities in the year 1921" (1958: 302).

Anyone who undertakes to criticize this theory is in danger of taking it more seriously than did Adams himself; he presumably did not expect any historian to accept this strange mixture of sense and nonsense as it stood, but rather wanted to stimulate speculation and research in a new direction.

In the third of his essays (1910), Adams made the second law of thermodynamics a very explicit basis for the tendency of history. He quoted Kelvin's conclusions on the dissipation of energy and Clausius' formulation of the second law using the concept of entropy; "to the vulgar and ignorant historian," Adams remarked, "it meant only that the ash-heap was constantly increasing in size" (1958: 138). If history was to concern itself (as Adams thought it should) with the future as well as the past, it can hardly ignore the latest result of physics which states that human society along with the physical universe must end in degradation and death, even though this prophecy is distasteful to the evolutionists who preach nothing but eternal progress. Evolution, together with Lyell's

The End

geological doctrine of uniformity, had already started to take the place of religious dogma, and "in a literary point of view the Victorian epoch rested largely,—perhaps chiefly,—on the faith that society had but to follow where science led . . . in order to attain perfection" (1958: 155). Nevertheless, geologists had been forced to take account of Kelvin's calculations of heat dissipation, which allowed only a few million years between times when the earth must be either too hot or too cold for habitation. The latest anthropological studies had not confirmed the notion that evolution has always been from lower to higher forms, even in the case of man. Adams could cite many examples from scientific literature and the popular press to illustrate trends toward biological and social degeneration, in confirmation of the second law of thermodynamics.

Another historian who took notice of the second law was Oswald Spengler. In his *Decline of the West* (1918), he said that among the various symbols of decline, "the most conspicuous is the notion of Entropy" (Spengler 1962: 216). He discussed the significance of the Second Law as indicating "the self-destruction of dynamic physics" *(ibid.)* by the introduction of *historical* concepts which are fundamentally opposed to the previously dominant spirit of mechanics. The actual irreversibility of natural processes, in contrast to the theoretical reversibility of processes in Newtonian mechanics, is linked to the statistical concept of "disorder" on the microscopic level. But

> Statistics belong, like chronology, to the domain of the organic, to fluctuating Life, to Destiny and Incident, and not to the world of Law and timeless causality. As everyone knows, statistics serve above all to characterize political and economic, that is, historical developments. And if, now, suddenly the contents of that field are supposed to be understood and understandable only statistically and under the aspect of Probability—instead of under that of the *a priori* exactitude which the Baroque thinkers unanimously demanded—what does it mean? It means that the object of understanding is ourselves. The Nature "known" in this wise is the Nature that we know by way of living experience, that we live in our-

selves. What theory asserts . . . —to wit, this ideal reversibility that never happens in actuality—represents a relic of the old severe intellectual form, the great Baroque tradition that had contrapuntal music for twin sister. But the resort to statistics shows that the force that tradition regulated and made effective is exhausted . . . the mythopoetic force of the Faustian soul is returning to its origins. Force, Will, has an aim, and where there is an aim there is for the inquiring eye an end. That which the perspective of oil-painting expressed by means of the vanishing point, the Baroque park by its *point de vue*, and analysis by the nth term of an infinite series—the conclusion, that is, of a willed directedness—assumes here the form of the concept. The Faust of the Second Part is dying, for he has reached his goal. What the myth of Götterdämmerung signified of old, the irreligious form of it, the theory of Entropy, signifies today (1962: 218–20).

Whatever one may think of Adams and Spengler as historians, one cannot ignore them as propounders of connections between science and culture, and as barometers of the cold winds blowing through the first decades of the twentieth century. Anyone sensitive to cyclic movements could see that neorealism had arrived on schedule: Planck's quantum theory (1900) revived Boltzmann's statistical physics, Einstein's theory of Brownian movement (1905) and its confirmation by Jean Perrin established the existence of atoms once and for all, an event that appeared to justify the revival of mechanistic theories in biology (Loeb 1915); in 1900 the rediscovery of Mendel's work of 1865 put genetics on a quantitative footing; artists and composers were quick to ditch romantic impressionism in favor of mathematical styles like cubism and the tone row. Progressive reform movements extended suffrage to women and seemed to be revitalizing democracy. But something happened to abort or distort the development of this promising new movement; it died in the 1920s without having realized its early promise or even giving birth to a vigorous successor. In the absence of any better ideas about where to go, neorealism turned to neoclassicism or Dada. Henry Adams' prediction that thought would come to the limit of its possibilities in 1921 has to be taken a little more seriously when one reads in the latest edition of the *Encyclopaedia Britannica* the pronouncement that the arts failed to move creatively beyond cubism and that all the so-called "new" movements of the twentieth century originated before 1914 (Barzun 1974). And Spengler's warning, that the resort to concepts of entropy and statistics implies the abandonment of causality and the incorporation of the observer into the object of understanding, now reads like an anticipation of the Copenhagen interpretation of quantum mechanics, the world view or perhaps world non-view that replaced neorealism in science.

Some cultural historians say that World War I shattered the natural tendencies of European culture; others claim there is something inherently artificial or sterile about this second swing away from romanticism, and that it lacked the dynamic creativity needed to establish a viable new style. But surely that is too harsh a judgment on the period that saw the flowering of most of the greatest names in twentieth-century science and culture—Niels Bohr, John Dewey, Albert Einstein, Sigmund Freud, Walter Gropius, James Joyce, Pablo Picasso, Bertrand Russell, Arnold Schoenberg, George Bernard Shaw, Igor Stravinsky, and Frank Lloyd Wright.

At one extreme, neorealism advocated revolution as an end in itself, cutting all ties with history. In 1923, Yevgeny Zamyatin contrasted the law of revolution—"red, fiery, deadly; but this death means the birth of new life"—with the law of entropy which is "cold, ice blue[,] . . . no longer deadly, but comfortable. The sun ages into a planet, convenient for highways, stores, beds, prostitutes, prisons[;] . . . if the planet is to be kindled into youth again, it must be set on fire, it must be thrown off the smooth highway of evolution" (Zamyatin 1970: 108).

Thermodynamics was now complete and could no longer inspire new philosophical debates. Walther Nernst pointed out in his lectures that the first law had three discoverers (Mayer, Joule, and Helmholtz), the second had two (Carnot and Clausius) but the third was the work of one person only—himself! Hence there could be no possible discoverer of a fourth law.

In fact Nernst's third law of thermodynamics—that all thermodynamic functions become zero at absolute zero temperature but that it is impossible ever to reach this temperature—would never have been established as an independent axiom had it not been shown to be harmonious with the new quantum theory. At the same time the role of thermodynamics as a model theory in physics was being taken over by Einstein's theory of relativity (1905). Not only did Einstein's theory generalize the principle of conservation of energy to include mass-energy transformations; it also introduced new criteria (invariance under certain coordinate transformation) that had to be satisfied by all other fundamental theories, just as the first and second laws of thermodynamics had previously been thought to impose restrictions on all other theories.

Scientific theories of the neorealist period often echoed those of the mid-nineteenth century realist period, but lacked the earlier confidence in the ultimate validity of a mechanistic approach. The emphasis was on mathematical form and structure, leaving aside questions of meaning and content; in mathematics itself Hilbert's "formalism" vanquished Brouwer's "intuitionism." In physics, the mechanists themselves were suspicious of the new trend; as Arthur Schuster wrote in 1904,

> Those who believe in the possibility of a mechanical conception of the universe, and are not willing to abandon the methods which, from the time of Galileo and Newton, have uniformly and exclusively led to success, must look with the gravest concern on a growing school of scientific thought which rests content with equations representing numerical relationships between different phenomena even though no precise meaning can be attached to the symbols used (Schuster 1924: vi).

The phenomenalism and irrationalism of the neoromantic period had been tamed and put into neat patterns. Even the term "degeneration" was introduced into physics, at first to describe the results of irreversible processes (Pfaundler 1904, Franklin 1910), but later simply to characterize a peculiar state of matter at low temperatures. Boltzmann had attacked empirio-criticism's predilection for nudity, but now with X rays and psychoanalysis scientists went beyond nudity to reveal the bones and bestial motives lying within our bodies and brains.

Sigmund Freud's career provides an interesting illustration of the intertwining of different cultural movements. The influence of romanticism (hearing a poem of Goethe) led him to Brücke, who was one of the physiologists who led the mechanist movement which wanted to destroy vitalism and put biology on a physico-chemical foundation. Freud was then led into a period of radical materialism in his early enthusiasm for "physicalistic physiology." His emphasis on the role of the unconscious and of irrational motives in human behavior makes him an important contributor to the neoromantic movement, but his mature version of psychoanalysis replaced the degeneration theory of neuroses. By 1921 Freud was willing to admit that, in contrast to the spiritualists, psychoanalysts are "at bottom incorrigible mechanists and materialists, even though they seek to avoid robbing the mind and spirit of their still unrecognized characteristics" (Freud 1955: 179).

While nowadays we usually think of psychoanalysis and behaviorism as opposite poles in psychology, there was considerable affinity between the two schools in earlier days. In the United States both were allied with the progressive movement in politics before World War I; social control based on rationalization of human behavior was a goal of political and scientific reformers. Frederick W. Taylor proclaimed, in his *Principles of Scientific Management*, that "in the past the man has been first; in the future the system must be first" (1911: 7). The search for efficient methods of organizing people was aided by J. B. Watson's behaviorist psychology (1913) and by Stern's invention of the "intelligence quotient" (1911), extensively developed by Lewis M. Terman and his colleagues at Stanford. Philosophers like Ralph Barton Perry who called themselves neorealists quickly endorsed the behaviorist method.

Artists and architects joined the psychologists and philosophers in embracing the new age of technology, putting aside earlier romantic distaste for machinery. The Italian futurists glorified the speeding automobile; the Deutsche Werkbund and the later Bauhaus promoted streamlined functional design. "Modernism" meant the collaboration of art and science to create the twentieth-century world; but the emphasis was on experiments in formal structure and rhythm rather than content. Lewis Mumford (1951) recalls some of the outstanding achievements of this movement:

> Duchamp's brilliant anticipation of stroboscopic photography in his painting of a Nude Descending a Staircase, Brancusi's transpositions of a bird in flight into a smoothly sinuous metal abstraction, Léger's inversion of the human body and other organic forms into cylinders and pistons, Gabo's translation of the exquisite materials and processes of modern technology into symbolic sculptures of metal, glass, and plastics, even the chaste if absurdly empty abstractions of Mondrian[;]. . . . the merely useful now became esthetically significant (Mumford 1951: 13).

That Freud as well as Watson could be considered neorealist suggests that the reality of neorealism was as likely to be inside one's mind as outside. For Picasso, the leading artist of the "cubist decade" (1905–14) if not of the entire century,

> being a realist did not mean reproducing the tangible world or making a sort of inventory of things, any more than it meant imitation, however cleverly transposed. What he aimed at doing was creating—with the materials appropriate to his art—a reality which would be the equivalent of reality as it is perceived, a reality in a sense more real than that of nature, a *mental object*, in other words a reality brought into being by the human mind (Elgar & Maillard 1957: 84).

That might also serve as a description of the "expressionist movement" which arose around 1920, looking backward to symbolism and ahead to surrealism.

Finding reality inside one's own mind and then expressing it in terms of the external world was also the procedure of Albert Einstein, who was turning away from the empiricism he inherited from Ernst Mach toward a platonic realism. Contrary to the traditional account of physics textbooks and of many philosophers of science, Einstein's special theory of relativity was not simply a response to the discrepancy between nineteenth-century ether theory and the Michelson-Morley experiment. Instead it was an attempt to resolve a puzzling asymmetry in Maxwell's electromagnetic theory, which described the electric current induced by a magnetic field in two different ways depending on which was supposed to be at rest and which in motion, even though the net result depended only on their relative motion. Einstein was apparently aware of the null result of Michelson's attempt to measure the motion of the earth through the ether only as one of several similar failures to determine such motion, and not the most striking one at that. After convincing himself of the validity of his special theory, Einstein plunged into the abstract world of non-Euclidean geometry and tensor calculus and, by demanding that the equations for gravitational fields satisfy a rigorous requirement of covariance under transformations of coordinate systems, arrived at his general theory of relativity in 1915. The confirmation of Einstein's prediction of the bending of light by the sun's gravitational field by a British expedition in 1919 was the greatest scientific triumph of neorealism. (It was also a belated triumph of internationalism in science, going against the nationalist passions aroused by the war.) But after 1920, Einstein became increasingly isolated from the rest of the physics community as he pursued his own search for absolutes while his colleagues were seduced by empiricism and indeterminism.

The slogan "everything is relative" is sometimes thought to represent Einstein's influence on culture, but it seems to me that Arnold Schoenberg's *Tonreihe* is the artistic innovation closest in spirit to relativity theory. Abandoning the domination of the musical composition by a single key, as Einstein had abandoned the domination of physical theory by a single absolute space, Schoenberg insisted that each of the 12 notes in the octave be treated on an equal basis, just as Einstein insisted that every coordinate system is as good as every other.

There is no question of a direct influence on culture of Einstein's special theory of relativity, published in 1905 but not accepted by most of the scientific community until the following decade. If we see Schoenberg abandoning tonality, Picasso and the Cubists abandoning perspective, and the mathematician Hermann Minkowski abandoning the distinction between space and time, all working toward a new way of representing spatial and temporal dimensions, and all presenting their results in a two-year period (1907–1908), we must either conclude that this is a meaningless coincidence or that all three were stimulated by the same *Zeitgeist*. Fifteen years later, when the public became aware that Einstein had done some amazing things with "non-Euclidean geometry," involving the bending of light and the curvature of space, the idea got about that one formal system, even though internally self-consistent and apparently plausible, could just as well be replaced by another one. At that point scientific naturalism became a threat to traditional moral codes; cultural relativism and pragmatism gave birth to "functionalism," the doctrine that scientists need only describe how things work in a particular society without worrying about how things *should* work.

Thus neorealism began with the proclamation of unique democratic formulas, and ended with a democracy of all formulas. The mathematical-egalitarian world view was obviously vulnerable to fantastic misuse by lesser intellects. The mania for quantitative research in the social sciences was a striking feature of the late neorealist period. Kelvin (1883) had shown the way:

> I often say that when you can measure what you are speaking about and express it in numbers you know something about it; but when you cannot measure it, when you cannot express it in numbers, your knowledge is of a meagre and unsatisfactory kind: it may be the beginning of knowledge, but you have scarcely, in your thoughts, advanced to the stage of Science, whatever the matter may be (Kelvin 1891: 80).

No one remembered that Kelvin's own passion for applying mathematics to everything had led him into a colossal blunder in geology (Chapter III); the social scientists assumed that the prestige of the physical scientists would rub off on them if they could only be sufficiently quantitative in their research, and they engraved Kelvin's slogan on their new building at Chicago (1929): "When you cannot measure your knowledge is meager and unsatisfactory."

The notion that quantitative measurement is the essence of sci-

The End

ence is a disastrous confusion of means and ends. It is true that the greatest scientists have often searched for mathematical harmony in nature, but that harmony is certainly not obvious in raw numerical data. And the great issues in the history of science have not ultimately been about numbers but about qualitative questions: *not* "what is the mechanical equivalent of heat?" but "is heat a substance or a form of motion?"; *not* "what is the maximum efficiency of a steam engine?" but "is there a fundamental irreversibility in nature?"; *not* "how fast does the earth move through the absolute space?" but "is there really any such thing as absolute space?" Kelvin was absolutely correct when he pointed out in 1870 that our ability to estimate *quantitatively* the size and mass of an atom makes an enormous difference to the answer we give to the *qualitative* question, "do atoms exist?" But he was merely expressing a personal opinion (though a widely held one) when he limited scientific knowledge to quantitative knowledge in the above quotation, just as he was expressing a personal opinion (no longer a widely held one) when he said in 1884, "I never satisfy myself until I can make a mechanical model of a thing. If I can make a mechanical model I can understand it" (Kelvin 1884: 270). And one of the theories which Kelvin rejected, on the grounds that he could not make a comprehensible mechanical model of it, was Maxwell's electromagnetic theory of light, now regarded as one of the greatest scientific accomplishments of the nineteenth century.

Though I have been a little hard on Lord Kelvin in this book, I confess a secret admiration for his brilliant inventiveness, his versatility, and the nineteenth-century spirit which he represents. What was that spirit? I think of it as a dogged persistence in trying to associate abstract ideas with reality. It is precisely because of the increasing specialization of the twentieth century, and the accumulation of so many facts and theories in all fields of study that one has no time to grasp one of them before another comes along, that we fail to understand how certain very simple ideas which are part of our common universe of discourse—analysis and synthesis, constancy and degeneration—could have had so much influence in the nineteenth century. The nineteenth century did not just talk about these ideas or weave them into grand philosophical and aesthetic theories, as previous centuries had done. Instead, it tried to put them into practice, to measure and test them with quantitative experiments, and to make them the basis of very specific theories in physics, biology, geology, history, literature, and art. It proved the unity of forces in nature, attempted to determine the power of prayer, established a tendency toward degeneration in the human race and in the universe, deduced the creation from the theory of heat conduction, converted the myth of eternal return into a theorem of mechanics, and very nearly settled the question of the existence of atoms.

The urge to quantify, to make things scientific, also produced much that we find objectionable in the nineteenth century: the crude realism of some of its art and literature, the naive scientism of its social and historical theory, and the vulgar materialism of its popular scientific writing. These are things that cannot be ignored if one wants to understand the general trends in scientific and cultural movements as well as the individual achievements of great men. In fact it is usually the second-raters who are more "characteristic" of their own times than the geniuses; the latter may be leaders of movements, but they are also likely to be complex personalities who don't quite "fit" the description of a romantic or a realist. Likewise one can learn quite a lot about European and American social history around 1900 by following the influence of the theory of biological degeneration, though that theory is hardly mentioned in most modern histories of biology.

The history of science is much more than the history of those discoveries and theories that fall in a direct line of development leading to the accepted modern version of the subject. Often it is the popularizer of science who has more impact on his age than the discoverer; and the reasons why scientists obstinately reject the "right" theory and accept the "wrong" one at a certain time may provide the essential clue to the understanding of the development of science. Clearly it is quite difficult for any one person to learn enough about both the science and culture of recent centuries to write this kind of history with much confidence. Nevertheless the attraction of the subject is irresistible, and one can always hope to learn something new from the protests of the experts upon whose domains he has encroached.

General Bibliography

Barzun, Jacques
   1974 "European culture since 1800," *Encyclopaedia Britannica*, 15th ed. Chicago: Encyclopaedia Britannica, Inc. *Macropaedia* 6:1066–81.

Brush, Stephen G.
   1974 "The Development of the Kinetic Theory of Gases, VIII. Randomness and Irreversibility," *Archive for History of Exact Sciences* 12: 1–88 (1974). Reprinted in *The Kind of Motion we call Heat*. Amsterdam: North-Holland Pub. Co., 1976, pp. 543–654.

Gillispie, C. C.
   1960 *The Edge of Objectivity, An Essay in the History of Scientific Ideas*. Princeton, N. J.: Princeton University Press.

Hallier, Ernst
   1889 *Kulturgeschichte des Neunzehnten Jahrhunderts in ihren Beziehungen zu der Entwickelung der Naturwissenschaften*. Stuttgart: Enke.

Holton, Gerald, and Brush, Stephen G.
   1973 *Introduction to Concepts and Theories in Physical Science*, 2d ed. Reading, Mass.: Addison-Wesley.

Mandelbaum, Maurice
   1971 *History, Man and Reason: A Study in 19th–Century Thought*. Baltimore: Johns Hopkins Press.

Merz, J. T.
   1904 *History of European Thought in the Nineteenth Century*, 4 vols. Edinburgh: Blackwood, 1904–14.

Olson, Richard, ed.
   1971 *Science as Metaphor: The Historical Role of Scientfic Theories in Forming Western Culture*. Belmont, Calif.: Wadsworth.

Opper, Jacob
   1973 *Science and the Arts. A Study in Relationships from 1600–1900.* Rutherford/Madison/Teaneck, N.J.: Fairleigh Dickinson University Press. Music and scientific-cultural movements (primarily Newtonian and Darwinian).

Priestley, J. B.
   1960 *Literature and Western Man.* New York: Harper.

Schneider, H. W.
   1946 *A History of American Philosophy.* New York: Columbia University Press.

Sypher, Wylie
   1960 *Rococo to Cubism in Art and Literature.* New York: Vintage.

Taton, René, ed.
   1965 *Science in the Nineteenth Century*, trans. from French by A. J. Pomerans. New York: Basic Books.

Chapter Bibliographies

*Chapter I**

Arnold, Matthew
  1869 *Culture and Anarchy*. London: Smith, Elder.
Bever, Thomas G., and Chiarello, Robert J.
  1974 "Cerebral dominance in musicians and nonmusicians," *Science* 185: 537–39.
Bocheński, I. M.
  1961 *Contemporary European Philosophy*. Berkeley: University of California Press.
Boring, E. G.
  1955 "Dual role of the *Zeitgeist* in Scientific Creativity," *Scientific Monthly* 80: 101–106.
Brinton, Crane
  1938 *The Anatomy of Revolution*. New York: Norton, 1938; rev. ed., Englewood Cliffs, N.J.: Prentice-Hall, 1952.
Brush, Stephen G.
  1970 "The Wave Theory of Heat: A forgotten stage in the transition from the caloric theory to Thermodynamics," *The British Journal for the History of Science* 5: 145–67 (1970). Reprinted in *The Kind of Motion we call Heat*, pp. 303–34. Amsterdam: North-Holland, 1976.
  1974 "Should the History of Science be rated X?" *Science* 183: 1164–72.
Butterfield, Herbert
  1931 *The Whig Interpretation of History*. London: Bell.

---

*The bibliography for each chapter includes sources pertaining to the subject, not all of which have been cited in the text.

Collingwood, R. G.
- 1927a "Oswald Spengler and the Theory of Historical Cycles," *Antiquity* 1: 311–25.
- 1927b "The Theory of Historical Cycles: II. Cycles and Progress," *Antiquity* 1: 435–46.

Crowe, Michael
- 1967 "Science a Century Ago." In *Science and Contemporary Society*, edited by F. J. Crosson, pp. 105–26. Notre Dame, Ind.: University of Notre Dame Press.

Cuneo, Ernest
- 1963 *Science and History*. New York: Duell, Sloan and Pearce. Theory of history based on conservation of energy.

Dicey, A. V.
- 1905 *Lectures on the Relation between law and public opinion in England, during the Nineteenth Century*, 2d ed. London: Macmillan.

Duncan, David
- 1908 *Life and Letters of Herbert Spencer*, Volume II. New York: Appleton.

Elkana, Yehuda
- 1974 *Discovery of the Conservation of Energy*. London: Hutchinson.

Gardiner, P., ed.
- 1959 *Theories of History*. New York: Macmillan, The Free Press of Glencoe.

Gottschalk, Louis
- 1972 "Three Generations: A plausible interpretation of the French *Philosophes?*" In *Irrationalism in the 18th Century*, edited by Harold E. Pagliaro, pp. 3-12. Cleveland: Press of Case Western Reserve University.

Hexter, Jack
- 1961 "The Historian and his Day." In *Reappraisals in History*, pp. 1–13. Evanston, Ill.: Northwestern University Press.

Hill, Christopher
- 1965 *Intellectual Origins of the English Revolution*. Oxford: Clarendon Press.

Holton, Gerald
- 1962 "Über die Hypothesen, welche der Naturwissenschaft zugrunde liegen," *Eranos-Jahrbuch* 31: 351–425.
- 1964 "Stil und Verwirklichung in der Physik," *Eranos-Jahrbuch* 33: 319–63.

1967 "The Thematic Imagination in Science." In *Science and Culture*, edited by G.Holton, pp. 88–108. Boston: Beacon.
Hughes, H. S.
 1952 *Oswald Spengler: A Critical Estimate*. New York: Charles Scribner's Sons.
Iggers, G. G.
 1958 "The idea of progress in recent philosophies of history." *Journal of Modern History* 30: 215–26.
Kampf, Alfred
 1948 *Die Revolte der Instinkte*. Berlin: Verlag Volk und Welt.
[Kelvin] Thomson, William
 1870 "On the Size of Atoms," *Nature* 1: 551–53.
Knowles, J. T.
 1869 "The Alternation of Science and Art in History," *Contemporary Review* 10: 285–95.
Kuhn, Thomas S.
 1962 *The Structure of Scientific Revolutions*. Chicago: University of Chicago Press.
 1968 "History of Science," *International Encyclopedia of the Social Sciences*. New York: Macmillan, Vol. 14: 74–83.
 1974 "Second Thoughts on Paradigms." In *The Structure of Scientific Theories*, edited by Frederick Suppe, pp. 459–82. Urbana: University of Illinois Press.
Leavis, F. R.
 1962 *Two Cultures? The Significance of C. P. Snow*. London: Chatto and Windus, 1962; New York: Pantheon, 1963.
Manuel, Frank
 1965 *Shapes of Philosophical History*. Stanford: Stanford University Press.
Martindale, Colin
 1975 *Romantic Progression. The Psychology of Literary History*. Washington, D. C.: Hemisphere/New York: Halsted (Wiley).
Mora, George
 1961 "Historiographic and cultural trends in psychiatry: A Survey," *Bulletin of the History of Medicine* 35: 26–36.
Parsons, Talcott
 1967 "Unity and Diversity in the Modern Intellectual Disciplines: The Role of the Social Sciences." In *Science and Culture*, edited by Gerald Holton, pp. 39–69. Boston: Beacon.

Popper, Karl R.
  1957 *The Poverty of Historicism.* London: Routledge & Kegan Paul.
Rainoff, T. J.
  1929 "Wave-like fluctuations of creative productivity in the development of West-European physics in the eighteenth and nineteenth centuries," *Isis* 12: 287–307.
Rand, Walter
  1971 "Eulerian and Lagrangian Analogy applied to the History of Science and Technology," *Actes du XIIIe Congrès International d'Histoire des Sciences, Moscow, 1971.* Moscow: Nauka, 1974, Vol. 1, pp. 176–82.
Rousseau, George S.
  1972 "Are there really men of both cultures?" *Dalhousie Review* 52: 351–72.
Schapiro, Meyer
  1953 "Style." In *Anthropology Today*, pp. 287–312. Chicago: University of Chicago Press.
Schneer, Cecil J.
  1969 *Mind and Matter.* New York: Grove. Chapter 11, "The idea of energy and the assault on materialism."
Schofield, Robert E.
  1970 *Mechanism and Materialism: British Natural Philosophy in an Age of Reason.* Princeton, N. J.: Princeton University Press.
Somervell, D. C.
  1929 *English Thought in the Nineteenth Century.* London: Longmans.
Sorokin, P. A.
  1937 *Social and Cultural Dynamics.* New York: American Book Co., Vol. II, Chapter 10, "Fluctuation of the linear, cyclical, and mixed conceptions of the cosmic, biological and sociocultural processes."
Suppe, Frederick
  1974 "The Search for Philosophic Understanding of Scientific Theories." In *The Structure of Scientific Theories*, edited by F. Suppe, pp. 1–232. Urbana: University of Illinois Press.
Toulmin, Stephen
  1972 *Human Understanding*, Vol. 1. Princeton, N.J.: Princeton University Press.

Truesdell, C.
  1968 *Essays in the History of Mechanics.* New York: Springer. See the review by S. G. Brush in *Isis* 61: 115–18 (1970).
Williams, Raymond
  1958 *Culture and Society 1780–1950.* New York: Columbia University Press.
Williams, W. Mattieu
  1870 *The Fuel of the Sun.* London: Simpkin, Marshall & Co.
Zagorin, P.
  1959 "Historical Knowledge: A review article on the philosophy of history," *Journal of Modern History* 31: 243–55.

Chapter II

Ackerknecht, E. H.
  1932 "Beiträge zur Geschichte der Medizinalreform von 1848," *Sudhoffs Archiv für Geschichte der Medizin* 25: 61–109, 113–83. Relation between politics and medicine.
Ashby, Eric
  1958 *Technology and the Academies. An Essay on Universities and the Scientific Revolution.* London: Macmillan, 1958; New York: St. Martin's, 1963. Interactions of the British universities with humanism and German science.
Barzun, J.
  1941 *Darwin, Marx, Wagner.* Boston: Little, Brown.
  1943 *Romanticism and the Modern Ego.* Boston: Little, Brown.
Benn, A. W.
  1906 *The History of English Rationalism in the Nineteenth Century.* London: Longmans, Green. See esp. Chaps. IX and XIV. Effect of David Strauss' *Life of Jesus* in accelerating the transition to realism in England; survey of realist philosophy.
Binkley, R. C.
  1935 *Realism and Nationalism, 1852–1871.* New York: Harper.
Bowle, John
  1954 *Politics and Opinion in the 19th Century: An Historical Introduction.* New York: Oxford University Press. Book I: The Political Thought of the Romantic Age. Book II: The Political Thought of the Age of Darwin.

Bozeman, T. D.
  1972 "Science and Nineteenth-Century American Culture: A Note on George H. Daniels' *Science in the Age of Jackson*," *Isis* 63: 397–400. On the influence of Scottish realism.
Brinton, Crane
  1953 "Something went wrong: Three Views of the Heritage of the Early Nineteenth Century," *Journal of the History of Ideas* 14: 457–62. Review of Baumgardt, Hayek, Viereck books.
Brush, Stephen G.
  1957 "The Development of the Kinetic Theory of Gases, I. Herapath," *Annals of Science* 13: 188–98.
  1958a "The Development of the Kinetic Theory of Gases, III. Clausius," *Annals of Science* 14: 185–96.
  1958b "The Development of the Kinetic Theory of Gases, IV. Maxwell," *Annals of Science* 14: 243–55.
  1963 "The Royal Society's First Rejection of the Kinetic Theory of Gases (1821): John Herapath versus Humphry Davy," *Notes and Records of the Royal Society of London* 18: 161–80.
Cannon, W. F.
  1962 "The Role of the Cambridge Movement in Early 19th Century Science," *Proceedings of the Tenth International Congress of the History of Science, Ithaca, 1962*, pp. 317–20. Paris: Hermann, 1964.
  1964 "The Normative Role of Science in Early Victorian Thought," *Journal of the History of Ideas* 25: 487–502. Darwin shattered the alliance between science and religion that had existed since Newton.
Clive, John
  1957 *Scotch Reviewers: The Edinburgh Review, 1802–1815*. London: Faber and Faber.
Commager, Henry Steele
  1960 *The Era of Reform*. Princeton: Van Nostrand, 1960; introduction reprinted in his *The Search for a Usable Past*. New York: Knopf, 1967, pp. 168–80. Paradox that romanticism can have opposite political influence in different countries.
Copland, Aaron
  1968 *The New Music 1900–1960*. New York: Norton.

Cranefield, P. F.
1957 "The Organic Physics of 1847 and the Biophysics of Today," *Journal of the History of Medicine* 12: 407–23.
Culotta, Charles A.
1974 "German biophysics, objective knowledge, and Romanticism," *Historical Studies in the Phyiscal Sciences* 4: 3–38.
Davie, G. E.
1961 *The Democratic Intellect; Scotland and Her Universities in the Nineteenth Century*. Edinburgh: Edinburgh University Press. Includes an account of the influences of mathematical education.
Dodd, George
1970 "Wordsworth and Hamilton," *Nature* 228: 1261–63.
Ellegård, Alvar
1957 "Darwinian theory and nineteenth-century philosophies of science," *Journal of the History of Ideas* 18: 362–93.
Eriksson, Gunnar
1969 *Romantikens världsbild speglad i 1800-talets Svenska vetenskap*. Stockholm: Wahlström & Widstrand.
Foote, G. A.
1951 "The Place of Science in the British Reform Movement, 1830–1850," *Isis* 42: 192–208.
Fullmer, J. Z.
1960 "The Poetry of Sir Humphry Davy," *Chymia* 6: 102–26. Davy's literary anticipation of the conservation of energy.
1962 "Humphry Davy's *Weltanschauung*," *Proceedings of the 10th International Congress of the History of Science, 1962*, pp. 325–28. Paris: Hermann, 1964.
Galaty, David H.
1974 "The philosophical basis of mid-19th century German reductionism." *Journal of the History of Medicine* 29: 295–316.
Galdston, I.
1956 "The Romantic period in Medicine," *Bulletin of the New York Academy of Science* 32: 346–62.
Glass, Bentley
1953 "The long neglect of a scientific discovery: Mendel's laws of inheritance." In *Studies in Intellectual History*, edited by G.

Boas et al., pp. 148–60. Baltimore: Johns Hopkins Press.

Gode-von Aesch, A. G. F.
 1941 *Natural Science in German Romanticism*. New York: Columbia University Press.

Goodstein, Judith Ronnie
 1969 *Sir Humphry Davy: Chemical Theory and the Nature of Matter*. Ph.D. Dissertation, University of Washington. See also the introductory note by T. L. Hankins *et al.* on the views of R. Siegfried and L. P. Williams concerning influences on Davy.

Gower, Barry
 1973 "Speculation in physics: The theory and Practice of *Naturphilosophie*," *Studies in the History and Philosophy of Science* 3: 301–56.

Grabo, Carl
 1939 "Science and the Romantic Movement," *Annals of Science* 4: 191–205.

Hall, E. W.
 1956 *Modern Science and Human Values: A Study in the History of Ideas*. Princeton, N.J.: Van Nostrand.

Hayek, F. A.
 1952 *The Counter-Revolution of Science: Studies on the Abuse of Reason*. New York: Macmillan, The Free Press of Glencoe.

Heimann, P.M.
 1974 "Helmholtz and Kant: The Metaphysical Foundations of *Über die Erhaltung der Kraft*." *Studies in History and Philosophy of Science* 5: 205–38.

Hennemann, Gerhard
 1959 *Naturphilosophie im 19. Jahrhundert*. Freiburg/München: Alber.
 1967 "Der Dänische Physiker Hans Christian Oersted und die Naturphilosophie der Romantik," *Philosophia Naturalis* 10: 112–22.

Hesse, Mary B.
 1961 *Forces and Fields: The Concept of Action at a Distance in the History of Physics*. London: Nelson.

Höffding, H.
 1900 *A History of Modern Philosophy: A Sketch of the History of Philosophy from the Close of the Renaissance to our own Day*.

London: Macmillan, Vol. 2.

Holton, Gerald

1962 "Über die Hypothesen, welche der Naturwissenschaft zugrunde liegen," *Eranos-Jahrbuch* 31: 351–425.

1964 "Presupposition in the Construction of Theories." In *Science as a Cultural Force*, edited by Harry Woolf, pp. 77–108. Baltimore: Johns Hopkins Press.

Hook, Sidney

1965 "Hegel Rehabilitated?" *Encounter* 24 (1): 53–58. Relation of his philosophical and political views.

Jaki, Stanley L.

1969 "Goethe and the physicists," *American Journal of Physics* 37: 195–203.

Jammer, Max

1975 *Concepts of Force*. Cambridge, Mass.: Harvard University Press.

Kargon, Robert

1964 "William Rowan Hamilton, Michael Faraday, and the revival of Boscovichean Atomism," *American Journal of Physics* 32: 792–95.

Knight, David

1967a "The Scientist as Sage," *Studies in Romanticism* 6: 65–88.

1967b "Steps toward a Dynamical Chemistry," *Ambix* 14: 179–97.

1970 "The Physical Sciences and the Romantic Movement," *History of Science* 9: 54–75.

Kuhn, T. S.

1959 "Energy Conservation as an Example of Simultaneous Discovery." In *Critical Problems in the History of Science*, edited by M. Clagett, pp. 321–56. Madison, Wisc.: University of Wisconsin Press.

Leonard, Neil

1966 "Edward MacDowell and the Realists," *American Quarterly* 18: 175–82.

Leverette, W. E., Jr.

1965 "E. L. Youmans' crusade for Scientific Autonomy and Respectability," *American Quarterly* 17: 12–32. On Youmans' use of his *Popular Scientific Monthly* to defend naturalism and mechanism.

Lilley, S.
  1949 "Social Aspects of the History of Science," *Archives Internationales d'Histoire des Sciences* 6: 376–443.
Lipman, T. O.
  1964 "Wöhler's preparation of urea and the fate of vitalism," *Journal of Chemical Education* 41: 452–58.
Lovejoy, A. O.
  1936 *The Great Chain of Being.* Cambridge, Mass.: Harvard University Press.
  1948 *Essays in the History of Ideas.* Baltimore: Johns Hopkins Press, Chapter X.
Mayer, J. Robert
  1842 "Bemerkungen über die Kräfte der unbelebten Natur," *Annalen der Chemie und Pharmacie* 42: 233–40 (1842). English translation in S. G. Brush, *Kinetic Theory*, vol. 1. Oxford: Pergamon Press, 1965. See also R. B. Lindsay, *Julius Robert Mayer, Prophet of Energy.* Oxford: Pergamon Press, 1973.
  1863 "Remarks on the Mechanical Equivalent of Heat," *Philosophical Magazine*, ser. 4, 25: 493–522. Includes his repudiation of *Naturphilosophie*.
Mead, G. H.
  1936 *Movements of Thought in the Nineteenth Century.* Chicago: University of Chicago Press.
Mencher, Samuel
  1964 "The influence of Romanticism on Nineteenth-Century British social work," *Social Service Review* 38: 174–90.
Mendelsohn, Everett
  1964 "The biological sciences in the nineteenth century: Some problems and sources," *History of Science* 3: 39–59.
  1965 "Physical models and physiological concepts: Explanation in Nineteenth-Century biology," *The British Journal for the History of Science* 2: 201–19.
Nordenskiöld, E.
  1928 *The History of Biology, A Survey* (translated from Swedish). New York: Knopf.
Oersted, H. C.
  1852 *The Soul in Nature.* London: Bohn.
Opper, Jacob

1973 *Science and the Arts: A Study in Relationships from 1600–1900.* Rutherford, Madison, Teaneck, N.J.: Fairleigh Dickinson University Press. Deals primarily with music.

Ostwald, Wilhelm
1911 "The biology of the Savant: A study in the Psychology of Personality," *Scientific American Supplement* 72: 169–71.

Persons, S.
1958 *American Minds; A History of Ideas.* New York: Holt.

Pritchard, Charles
1869 "Spectrum Analysis," *Contemporary Review* 11: 481–90.

Rádl, E.
1930 *The History of Biological Theories* (translated from German). Oxford: Oxford University Press. Deals mainly with Darwinism.

Riley, W.
1923 *American Thought: From Puritanism to Pragmatism and Beyond.* New York: Holt.

Schnabel, F.
1934 *Deutsche Geschichte im Neunzehnten Jahrhundert. Dritter Band: Erfahrungswissenschaften und Technik.* Freiburg: Herder Verlag, 1934; zweite Auflage, 1950.

Shaffer, Elinor S.
1974 "Coleridge and natural philosophy: A review of recent literary and historical research," *History of Science* 12: 284–98. Rejects the thesis of L. P. Williams on transfer of nature philosophy to England.

Shryock, R. H.
1947 *The Development of Modern Medicine.* New York: Knopf.

Siegfried, Robert
1967 "Boscovich and Davy: Some cautionary remarks," *Isis* 58: 236–38. Critique of Williams (1964).

Snelders, H. A. M.
1970 "Romanticism and *Naturphilosophie* and the inorganic natural sciences, 1797–1840. An introductory survey," *Studies in Romanticism* 9: 193–215.
1971 "J. S. C. Schweigger: His romanticism and his crystal electrical theory of matter," *Isis* 62: 328–38.
1973 "Numerology in German Romanticism and 'Natur-

philosophie,' " *Janus* 60: 25–40.

Spencer. J. B.
1967 "Boscovich's Theory and its Relation to Faraday's Researches: An Analytic Approach," *Archive for History of Exact Sciences* 4: 184–202.

Stauffer, R. C.
1953 "Persistent errors regarding Oersted's discovery of electromagnetism," *Isis* 44: 307–10.
1957 "Speculation and experiment in the background of Oersted's discovery of electromagnetism," *Isis* 48: 33–50.

Temkin, Owsei
1946a "The Philosophical Background of Magendie's Physiology," *Bulletin of the History of Medicine* 20: 10–35.
1946b "Materialism in French and German Physiology of the Early Nineteenth Century," *Bulletin of the History of Medicine* 20: 322–27.
1963 "The Fielding H. Garrison lecture: Basic science, medicine, and the Romantic era," *Bulletin of the History of Medicine* 37: 97–129.

Temple, George
1954 *Classic and Romantic in Natural Philosophy.* Oxford: Clarendon.

Thomas, R. Hinton
1951 *Liberalism, Nationalism, and the German intellectuals (1822–1847). An Analysis of the Academic and Scientific Conferences of the Period.* Cambridge, England: Heffer.

Thomson, David
1955 "Scientific thought and revolutionary movements," *Impacts of Science on Society* 6: 3–29.

Tindall, W. Y.
1956 *Forces in Modern British Literature, 1885–1956.* New York: Vintage Books.

Wachsmuth, Bruno
1939 "Romantische Naturwissenschaft—ihre Grundzüge und ihr Erlöschen im 19. Jahrhundert," *Klinische Wochenschrift* 18: 998–1004.

Weiss, P. A.
1964 "The emergence of scientific thought in the eighteenth cen-

tury: Some improvisations," *The Graduate Journal* 6: 377–94.

Wetzels, Walter D.
  1971 "Aspects of natural science in German Romanticism," *Studies in Romanticism* 10: 44–59.

Whyte, Lancelot L., ed.
  1961 *Roger Joseph Boscovich, S. J., F. R. S., 1711–1787, Studies of his Life and Work on the 250th Anniversary of his Birth.* London: George Allen and Unwin.

Wiener, Norbert
  1951 "Pure and applied mathematics,." In *Structure, Method and Meaning: Essays in honor of Henry M. Sheffer,* ed. Paul Henle, Horace M. Kallen, and Susanne K. Langer, pp. 91–98. New York: Liberal Arts Press.

Wilde, Oscar
  1891 *The Picture of Dorian Gray.* London: Ward Lock.

Williams, L. Pearce
  1964 *Michael Faraday, A Biography.* London: Chapman and Hall.
  1973 "Kant, *Naturphilosophie* and Scientific Method." In *Foundations of Scientific Method: The Nineteenth Century,* edited by R. N. Giere and R. S. Westfall, pp. 3–22. Bloomington: Indiana University Press.

Chapter III

Albritton, C. C., ed.
  1975 *Philosophy of Geohistory: 1785–1970.* Stroudsburg, Pa.: Dowden, Hutchinson & Ross.

Bevington, M. M.
  1961 *The Saturday Review, 1855–1868.* New York: Columbia University Press. See pp. 282–84 on review of *Origin of Species* that led Darwin to retract his estimate of 300 million years for a geological process.

Boltwood, B. B.
  1907 "On the ultimate disintegration products of the radioactive elements, Part II. The disintegration products of uranium," *American Journal of Science* 23: 77–80, 86–88.

Burchfield, Joe D.

1974 "Darwin and the Dilemma of Geological Time," *Isis* 65: 301–21.
1975 *Lord Kelvin and the Age of the Earth*. New York: Science History Pubs.

Burke, John
1974 "The earth's central heat: from Fourier to Kelvin," *Actes du XIII$_e$ Congrès International d'Histoire des Sciences, Moscow, 1971*. Section VIII: 118–23. Moscow: Editions "Nauka."

Carnot, Sadi
1824 *Réflexions sur la Puissance Motrice du Feu et sur les machines propres a devélopper cette puissance*. Paris: Bachelier. For English translation see Mendoza (1960).

Clausius, R.
1865 "Über verschiedene für die Anwendung bequeme Formen der Hauptgleichungen der mechanischen Wärmetheorie," *Annalen der Physik*, ser. 2, 125: 353–400 (1865). English translation in *The Mechanical Theory of Heat*. London: Macmillan, 1879.
1867 "On the second fundamental theorem of the mechanical theory of heat," *Philosophical Magazine*, ser. 4, 35: 405–19 (1868).

Cope, E. D.
1867 "The progress of discovery of the laws of evolution," *American Naturalist* 10: 218–27. Includes discussion of Haeckel's theory.

[Darwin, Charles]
1903 *More Letters of Charles Darwin*. New York: Appleton.
1959 *Variorum Text of the Origin of Species*, ed. M. Peckham. Philadelphia: University of Pennsylvania Press. Indicates changes from one edition to another.

Darwin, Emma
1915 *A Century of Family Letters*. New York: Appleton.

Eiseley, Loren
1958 *Darwin's Century*. Garden City: Doubleday, Chapter IX.

Eve, A. S.
1939 *Rutherford*. Cambridge: Cambridge University Press.

Fourier, J. B. J.
1807 "Théorie de la propagation de la chaleur dans les solides."

In *Joseph Fourier 1768–1830*, ed. I. Grattan-Guinness and J. R. Ravetz. pp. 33–440. Cambridge, Mass.: MIT Press, 1972,
1822 *Théorie Analytique de la Chaleur*. Paris: Didot, 1822. Reprinted in *Oeuvres de Fourier*, edited by G. Darboux, Vol. 1. Paris: Gauthier-Villars, 1888. English translation by A. Freeman, *The Analytical Theory of Heat*. New York: Dover, 1955 reprint.

Geikie, A.
1871 "On modern denudation," *Transactions of the Geological Society of Glasgow* 3: 153–90.
1899 "Presidential Address to the Geological Section." *Report of the British Association Meeting*, 718–30.

Gillispie, C. C.
1951 *Genesis and Geology*. Cambridge, Mass.: Harvard University Press.

Greenough, George
1834 "Anniversary Address to the Geological Society of London, 1834," *Proceedings of the Geological Society of London* 2: 42–70 (1838).

Haber, Francis C.
1959 *The Age of the World: Moses to Darwin*. Baltimore: Johns Hopkins Press.

Helmholtz, H. von
1854 "Ueber die Wechselwirkung der Naturkräfte und die darauf bezüglich neuesten Ermittelungen der Physik." Königsberg: Gräfe & Unzer. English translation by J. Tyndall, "On the Interaction of Natural Forces," *Philosophical Magazine* 11: 489–518 (1856). Reprinted in *Popular Scientific Lectures*, edited by M. Kline, pp. 59–92. New York: Dover, 1962.

Himstedt, F.
1904 "Über die radioaktive Emanation der Wasser- und Ölquellen," *Physikalische Zeitschrift* 5: 210–13.

Huxley, T. H.
1869 "Anniversary Address of the President," *Quarterly Journal of the Geological Society of London* 25: xxviii–liii.
1894 *Science and Hebrew Tradition*. New York: Appleton.
1896 *Discourses—Biological and Geological*. New York: Appleton.

Kelland, Philip
1837 *Theory of Heat*. Cambridge, Eng.: Dighton.

[Kelvin] Thomson, William
- 1841 "On Fourier's expansions of functions in trigonometrical series," *Cambridge Mathematical Journal* 2: 258–62.
- 1842 "On the linear motion of heat," *Cambridge Mathematical Journal* 3: 170–74, 206–11.
- 1852 "On a Universal Tendency in Nature to the Dissipation of Mechanical Energy," *Philosophical Magazine*, ser. 4, 4: 304–306.
- 1862a "On the age of the Sun's heat," *Macmillan's Magazine* 5: 388–93; *Philosophical Magazine*, ser. 4, 23: 158–60.
- 1862b "On the secular cooling of the earth," *Transactions of the Royal Society of Edinburgh* 23:157–70 (1862); *Philosophical Magazine*, ser. 4, 25: 1–14 (1863).
- 1865 "The 'Doctrine of Uniformity' in Geology briefly refuted," *Proceedings of the Royal Society of Edinburgh* 5: 512–13.
- 1868 "On geological time," *Transactions of the Glasgow Geological Society* 3: 1–28.
- 1869 "Of Geological Dynamics," *Transactions of the Geological Society of Glasgow* 3: 215–38 (1871).
- 1871 "Address to the British Association Meeting at Edinburgh," *Report of the British Association Meeting* 41: lxxxiv–cv; *Nature* 4: 262–70.
- 1882 *Mathematical and Physical Papers*. Cambridge: Cambridge University Press, 1882–1911.
- 1892 "On the dissipation of energy," *Fortnightly Review*, ser. 2, 51: 313–21.
- 1894 *Popular Lectures and Addresses*. London: Macmillan.
- 1897 "The Age of the Earth as an Abode fitted for Life," *Annual Report of the Board of Regents of the Smithsonian Institution, July 1897*, pp. 337–57. Washington, D.C.: Government Printing Office (1898) Reprinted from *Victoria Institute Transactions*.

King, Clarence
- 1877 "Catastrophism and evolution," *American Naturalist* 11: 449–70.

LeConte, Joseph
- 1877 "On critical periods in the history of the earth, and their relation to evolution; on the Quarternary as such a period,"

American Naturalist 11: 540–57. Includes remarks on tendency toward stability in organic life.

Liebenow, C. H.
  1904 "Notiz über die Radiummenge der Erde," *Physikalische Zeitschrift* 5: 625–26.

Lyell, Charles
  1830 *Principles of Geology.* London: Murray, 1830–33.

Marchant, James
  1916 *Alfred Russel Wallace: Letters and Reminiscences.* New York: Harper.

Mendoza, E., ed.
  1960 *Reflections on the Motive Power of Fire by Sadi Carnot and other Papers on the Second Law of Thermodynamics by E. Clapeyron and R. Clausius.* New York: Dover.

Pfeifer, E. J.
  1965 "The genesis of American Neo-Lamarckism," *Isis* 56: 156–67.

Playfair, John
  1802 *Illustrations of the Huttonian Theory of the Earth.* Edinburgh: Cadell & Davies.

Rudwick, Martin J. S.
  1974 "Poulett Scrope on the volcanoes of Auvergne: Lyellian time and political economy," *British Journal for the History of Science* 7: 205–42.

Sharlin, H. I.
  1972 "On being scientific: A critique of evolutionary geology and biology in the Nineteenth Century," *Annals of Science* 29: 271–85.

Strutt, R. J.
  1905 "On the Radio-active Minerals," *Proceedings of the Royal Society of London* A76: 88–101.
  1906 "On the distribution of radium in the earth's crust, and on the earth's internal heat," *Proceedings of the Royal Society of London* A77: 472–85.

Tait, P. G.
  1871 "Address to the Mathematical and Physical Section," *Report of the British Association Meeting* 41: 1–8.

[Tait, P. G.]

1869 "Geological time," *North British Review* 50: 406–39.
Thompson, S. P.
  1910 *The Life of William Thomson, Baron Kelvin of Largs*. London: Macmillan, Vol. I, Chapter I.
Thomson, J. J.
  1937 *Recollections and Reflections*. New York: Macmillan.
Thomson, William
  See Kelvin.
Toulmin, Stephen, and Goodfield, June
  1965 *The Discovery of Time*. New York: Harper & Row.
Waterston, J. J.
  1853 "On Dynamical Sequences in Kosmos [presented at the 23rd meeting of the British Association for the Advancement of Science]," *Athenaeum* [volume for July-December 1853]: 1099–1100.
Wilkins, Thurman
  1958 *Clarence King*. New York: Macmillan
Wilson, David B.
  1974 "Kelvin's Scientific Realism: the Theological Context," *Philosophical Journal* 11: 41–60.
Wolf, A.
  1952 *A History of Science, Technology, and Philosophy in the 18th Century*, 2d ed. New York: Macmillan, Vol. I, Chapter XV.

Chapter IV

Boas, Franz
  1887 "The Study of Geography," *Science* 9: 137–41.
Brush, Stephen G.
  1973 "The Development of the Kinetic Theory of Gases. VII. Heat Conduction and the Stefan-Boltzmann Law," *Archive for History of Exact Sciences* 11: 38–96 (1973). Reprinted in *The Kind of Motion we call Heat*, pp. 469–542. Amsterdam: North-Holland, 1976.
Comte, August
  1830 *Cours de Philosophie Positive*. Paris: Bachelier, 1830–42.
Crane, H. Richard

1971 "Opportunities in geophysics," *Physics Today* 24 (2) (February): 23–26.
Crosland, M. P.
  1961 "The origins of Gay-Lussac's law of combining volumes," *Annals of Science* 17: 1–26 (1963).
Darwin, G. H.
  1879 "On the bodily tides of viscous and semi-elastic spheroids, and on the ocean tides upon a yielding nucleus," *Philosophical Transactions of the Royal Society of London* 170: 1–35.
  1898 *The Tides and Kindred Phenomena in the Solar System.* Boston: Houghton, Mifflin.
Dyson, F. J.
  1970 "The Future of Physics," *Physics Today* 23 (9) (September): 23–28.
Ehrenfest, P.
  1923 "Ein alter Trugschluss betreffs des Wärmegleichgewichtes eines Gases im Schwerefeld," *Zeitschrift für Physik* 17: 421–22.
Fourier, J. B. J.
  1827 "Mémoire sur les températures du globe terrestre et des espaces planétaires," *Mémoires de l'Académie Royale des Sciences et de l'Institut de France* 7: 570–604.
  1890 *Oeuvres de Fourier*, edited by G. Darboux. Volume 2. Paris: Gauthier-Villars,
Garber, Elizabeth
  1976 "Thermodynamics and Meteorology (1850–1900)," *Annals of Science* 33:51–65.
Geikie, Archibald
  1903 *Textbook of Geology*, 4th ed. London: Macmillan.
Gillmor, C. S.
  1975 "The place of the geophysical sciences in Nineteenth Century Natural Philosophy," *EOS* 56: 4–7.
Herapath, John
  1826 "Sir H. Davy and Mr. Herapath," *Times* (London), January 10, p. 3.
  1836 "Fall of temperature in ascending the atmosphere," *Railway Magazine* 1: 19–21.
  1847 *Mathematical Physics*. London: Whittaker and Co., and

Herapath's Railway Journal Office, 1847. Reprinted with introduction and bibliography by S. G. Brush, New York: Johnson Reprint Corp., 1972.

Hopkins, William
  1839 "On the phenomena of precession and nutation, assuming the fluidity of the interior of the earth," *Philosophical Transactions of the Royal Society of London* 129: 381–423.

Jeffreys, Harold
  1973 "Developments in Geophysics," *Annual Review of Earth and Planetary Sciences* 1: 1–13. Remarks on Kelvin vs. Huxley, and his own rejection of continental drift.

[Kelvin] Thomson, William
  1862 "On the rigidity of the earth," *Proceedings of the Glasgow Philosophical Society* 5: 169–70; *Philosophical Transactions of the Royal Society of London* 153: 573–82 (1863–64).
  1872 "The rigidity of the earth," *Nature* 5: 223–24, 257–59.

Knopoff, L.
  1972 "Significance and achievements of the upper mantle project," *ICSU Bulletin* 27 (September): 3–7.

Knott, Cargill G.
  1899 "The propagation of earthquake vibrations through the earth," *Proceedings of the Royal Society of Edinburgh* 22: 573–85.

Mackin, J. Hoover
  1963 "Rational and empirical methods of investigation in geology." In *The Fabric of Geology,* ed. C. C. Albritton. San Francisco: Freeman, Cooper, pp. 135–63. Remarks on quantification, the generation gap, etc.

Manley, Gordon
  1968 "Dalton's Accomplishments in Meteorology." In *John Dalton & the Progress of Science*, edited by D. S. L. Cardwell, pp. 14–158. New York: Barnes & Noble.

Maxwell, James Clerk
  1856 "On the stability of the motion of Saturn's Rings." In *The Scientific Papers of James Clerk Maxwell,* edited by W. D. Niven. Vol. 1, pp. 288–376. Cambridge: Cambridge University Press, 1890. (Adams Prize Essay for 1856)

Menard, Henry

1971 *Science: Growth and Change*. Cambridge, Mass.: Harvard University Press.

Oldham, R. D.
   1923 *Geographical Journal* 61: 188–90. Discussion remarks.

Physics Survey Committee of the National Research Council
   1966 *Physics: Survey and Outlook*. Washington, D.C.: National Academy of Sciences.
   1972 *Physics in Perspective*. Vol. 1. Washington, D.C.: National Academy of Sciences.

Planck, Max
   1909 "Die Einheit des Physikalischen Weltbildes," *Physikalische Zeitschrift* 19: 62–75. English translation in Planck (1960: 1–26).
   1960 *A Survey of Physical Theory*. Translated by R. Jones and D. H. Williams. New York: Dover.

Schneider, I.
   1974 "Clausius' erste Anwendung der Wahrscheinlichkeitsrechnung im Rahmen der atmosphärischen Lichtstreuung," *Archive for History of Exact Sciences* 14: 143–58.

Shaler, N. S.
   1896 "Relations of Geologic Science to Education," *Bulletin of the American Geological Society* 7: x, 315–26.

Swinton, W. E.
   1975 "The Relation of Geology to other Sciences," *Journal of the History of Ideas* 35: 729–38.

Thackray, Arnold
   1972 *John Dalton: Critical Assessments of His Life and Science*, pp. 64–88. Cambridge, Mass.: Harvard University Press.

Waterston, J. J.
   1893 "On the physics of media that are composed of free and perfectly elastic molecules in a state of motion," *Philosophical Transactions of the Royal Society of London* 183A: 5–79. Read 11 December 1845.

Weber, Max
   1904 " 'Objectivity' in Social Science and Social Policy," translated from *Archiv für Sozialwissenschaft und Socialpolitik* (1904). In *The Methodology of the Social Sciences*, pp. 50–112. New York: Free Press, 1949. Influence of the idea that laws are

more important than events in science.

Wegener, Alfred
1915 *Die Entstehung der Kontinente und Ozeane.* Braunschweig: Vieweg.
1927 "Die geophysikalischen Grundlagen der Theorie der Kontinentverschiebung," *Scientia* 41: 103–16. Translation of extract in Foreword to Wegener (1966).
1966 *The Origin of Continents and Oceans.* Translated from the 4th revised German ed. by John Biram. New York: Dover.

Wilson, Fred L.
1970 Review of Middleton, *Invention of the Meteorological Instruments, Physics Today* 23 (9) (September): 53.

Wilson, Leonard G.
1969 "The intellectual background to Charles Lyell's *Principles of Geology,* 1830–1833." In *Toward a History of Geology,* edited by C. J. Schneer, pp. 426–43. Cambridge, Mass.: MIT Press.

Chapter V

Andler, C.
1958 *Nietzsche, sa Vie et sa Pensée.* Paris: Gallimard. Vol. 4; Livre 2, Ch. I and Livre 3, Ch. I.

Antoine, Jean-Claude
1948 "L'Eternel Retour de l'histoire deviendra-t-il objet de science?" *Critique* (Paris) 27 (August): 723–38.

Becker, Oskar
1936 "Nietzsches Beweise für seine Lehre von der ewigen Wiederkunft," *Blattern für Deutsche Philosophie* 9: 368–87.

Bernfeld, S., and Feitelberg, S.
1930 "Der Entropiesatz und der Todestrieb," *Imago* 16: 187ff. English translation in *International Journal of Psychoanalysis* 12: 61–81 (1931). See critique by Kapp.

Blanqui, A.
1872 "La Cosmogonie de Laplace—les comètes," *Revue Scientifique de la France* 2: 797–803.

Boltzmann, Ludwig
1872 "Weitere Studien über das Warmegleichgewicht unter

Gasmolekülen," *Sitzungsberichte der kaiserlichen Akademie der Wissenschaften in Wien*, Abt. 2, 66: 275–370. English translation in Brush (1966).

1877 "Ueber die Beziehung eines allgemeine mechanischen Satzes zum zweiten Hamptsatze der Wärmetheorie," *Sitzungsberichte der kaiserlichen Akademie der Wissenschaften in Wien*, Abt. 2, 75: 67–73. English translation in Brush (1966).

1895 "On certain Questions of the Theory of Gases," *Nature* 51: 413–15, 581.

1896 "Entgegnung auf die wärmetheoretischen Betrachtungen des Hrn. Zermelo," *Annalen der Physik*, series 3, 57: 773–84. English translation in Brush (1966).

1896 *Vorlesungen über Gastheorie*. Leipzig: Barth, 1896–98. English translation with introduction and bibliography by S. G. Brush, *Lectures on Gas Theory*. Berkeley: University of California Press, 1964.

1897a "Zu Hrn. Zermelo's Abhandlung über die mechanische Erklärung irreversibler Vorgänge," *Annalen der Physik*, series 3, 60: 392–98. English translation in Brush (1966).

1897b "Ueber irreversible Strahlungsvorgänge." *Sitzungsberichte der Königlich Preussischen Akademie der Wissenschaften, Physikalisch-Mathematische Klasse, Berlin*, 660–62, 1016–18.

1898 "Ueber vermeintlich irreversible Strahlungsvorgänge." *Sitzungsberichte der Königlich Preussischen Akademie der Wissenschaften, Physikalisch-Mathematische Klasse, Berlin*, 182–87.

Brunhes, Bernard
1908 *La Dégradation de L'Énergie*. Paris: Flammarion.

Brush, Stephen G.
1966 *Kinetic Theory*, volume 2, *Irreversible Processes*. Oxford: Pergamon Press.
1976 "Irreversibility and Indeterminism: Fourier to Heisenberg," *Journal of the History of Ideas* 37: 603–30.

Bryan, G. H.
1891 "Researches related to the Connection of the Second Law with Mechanical Principles; the laws of Distribution of Energy and their Limitations," *Report of the British Association Meeting* 61: 85–122 (1891); 64: 64–106 (1894).

Burbury, S. H.

1894 "Boltzmann's Minimum Function," *Nature* 51; 78, 320 (1894); 52: 104–105 (1895).

Čapek, Milic
1960 "The theory of eternal recurrence in modern philosophy of science, with special reference to C. S. Peirce," *Journal of Philosophy* 57: 289–96.
1961 *The Philosophical Impact of Contemporary Physics*. Princeton, N.J.: Van Nostrand, Chapter VIII.

Carathéodory, Constantin
1919 "Über den Wiederkehrsatz von Poincaré," *Sitzungsberichte der Preussischen Akademie der Wissenschaften, Berlin* 34: 579–84.

Clausius, R.
1868 "On the Second Fundamental Theorem of the Mechanical Theory of Heat," *Philsophical Magazine*, series 4, 35: 405–19. Translated from the author's German text of a lecture given in September 1867.

Cocke, W. J.
1967 "Statistical time symmetry and two-time boundary conditions in physics and cosmology," *Physical Reivew*, series 2, 160: 1165–70.

Daub, Edward E.
1970 "Maxwell's Demon," *Studies in History and Philosophy of Science* 1: 213–27.

Dauvillier, A.
1963 *Les Hypothèses Cosmogoniques: Théories des Cycles Cosmiques et des Planetes Jumelles*. Paris: Masson.

Delevsky, Jacques
1945 "Note sur la possibilité des répétitions cosmologiques," *Isis* 36, 19–21.
1946 "L'idée du cycle éternel dans l'histoire du monde." In *Studies and Essays in the History of Science and Learning offered in Homage to George Sarton on the Occasion of his Sixtieth Birthday, 31 August 1944*. New York: Henry Schuman, pp. 375–402.

Duncan, David
1908 *Life and Letters of Herbert Spencer*. New York: Appleton.

Eddington, A. S.
1928 *The Nature of the Physical World*. Cambridge: Cambridge

University Press.
Ehrenfest, Paul and Tatiana
1912 "Begriffliche Grundlagen der statistischen Auffassung in der Mechanik," *Encyklopädie der mathematischen Wissenschaften*, Bd. IV, Teil IV, Art. 32. Leipzig: Teubner. English translation by M. J. Moravcsik, *The Conceptual Foundations of the Statistical Approach in Mechanics*. Ithaca: Cornell University Press, 1959.
Eliade, Mircea
1955 *The Myth of the Eternal Return*. New York: Pantheon.
Fechner, Gustav
1873 *Einige Ideen zur Schöpfungs- und Entwickelungsgeschichte der Organismen*. Leipzig: Breitkopf & Härtel.
Fiske, John
1874 *Outlines of Cosmic Philosophy*, Volume II. Boston: Houghton Mifflin, 1874, 1902. Critique of Comte and exposition of Spencer's philosophy.
Flammarion, Camille
1891 "The last days of the earth," *Contemporary Review* 59: 558–69. The heat death.
1894 *La Fin du Monde*. Paris: Flammarion.
Flugel, J. C.
1955 *Studies in Feeling and Desire*. London: Duckworth, Chapter IV. Freud's death instinct, Fechner on stability, Spencer on equilibrium.
Freud, Sigmund
1920 *Jenseits des Lustprinzips*. Leipzig: Internationaler Psychoanalytischer Verlag.
Gould, Stephen Jay
1970 "Dollo on Dollo's law: irreversibility and the status of evolutionary laws," *Journal of the History of Biology* 3: 189–212.
Harris, Frank
1963 *My Life and Loves*. New York: Grove Press, Vol. 3, Chap. 7. On his meeting with Kelvin.
Hartmann, Eduard von
1902 *Die Weltanschauung der modernen Physik*. Leipzig: Haacke.
Heimann, P. M.

1972 "The *Unseen Universe:* Physics and the philosophy of nature in Victorian Britain," *British Journal for the History of Science* 6:73–79.

Hiebert, Erwin
1966 "The uses and abuses of thermodynamics in religion," *Daedalus* 95: 1046–80.
1967 "Thermodynamics and religion: A historical appraisal." In *Science and Contemporary Society*, edited by F. J. Crosson, pp. 57–104. Notre Dame, Ind.: University of Notre Dame Press.
1968 *The Conception of Thermodynamics in the Scientific Thought of Mach and Planck.* Freiburg: Ernst-Mach Institut.

Hollingdale, R. J.
1965 *Nietzsche: The Man and his Philosophy.* London: Routledge and Kegan Paul, Appendix II: "Three Objections to the Recurrence Conflict with the Second Law."

Jaki, Stanley L.
1974 *Science and Creation: From Eternal Cycles to an Oscillating Universe.* New York: Science History Publications.

Jeans, J. H.
1929 *The Universe around Us.* Cambridge: Cambridge University Press.
1933 *The New Background of Science.* Cambridge: Cambridge University Press.

Kapp, R. O.
1931 "Comments on Bernfeld's and Feitelberg's 'The Principle of Entropy and the Death Instinct,'" *International Journal of Psychoanalysis* 12: 82–86. There is no entropy concept in psychology.

Kaufmann, Walter
1950 *Nietzsche: Philosopher, Psychologist, Antichrist.* Princeton: Princeton University Press, Chapter 11.

[Kelvin] Thomson, William
1874 "The Kinetic Theory of the Dissipation of Energy," *Proceedings of the Royal Society of Edinburgh* 8: 325–34. Reprinted in S. G. Brush (1966).
1894 *Popular Lectures and Addresses,* Volume 2. London: Macmillan.

Kennedy, E. S.

1964 "Ramifications of the World-Year concept in Islamic astrology." *Proceedings of the Tenth International Congress on History of Science, Ithaca, 1962.* Paris: Hermann, pp. 23–45.

Klein, Martin J.
1970 "Maxwell, his Demon, and the Second Law of Thermodynamics," *American Scientist* 58: 84–97.

Knott, C. G.
1911 *Life and Scientific Work of Peter Guthrie Tait.* Cambridge: Cambridge University Press.

Lalande, André
1899 *La Dissolution opposée à l'Évolution dans les Sciences Physiques et Morales.* Paris: Alcan.

Loschmidt, Josef
1876 "Über den Zustand des Wärmegleichgewichts eines Systems von Körpern mit Rucksicht auf die Schwerkraft," *Sitzungsberichte der kaiserliche Akademie der Wissenschaften in Wien*, Abt. 2, 73: 128–42, 366–72 (1876); 75: 287–98 (1877); 76: 209–25 (1878).

Löwith, K.
1935 *Nietzsche's Philosophie der ewigen Widerkunft des Gleichen.* Berlin: Verlag Die Runde.
1949 "Nietzsche's revival of the doctrine of eternal recurrence." In *Meaning in History: The Theological Implications of the Philosophy of History*, pp. 214–22. Chicago: University of Chicago Press.

Mach, Ernst
1894 "On the Principle of the Conservation of Energy," *The Monist* 5: 22–54. Reprinted in *Popular Scientific Lectures.* La Salle, Ill.: Open Court. 5th ed., 1943.

Maxwell, James Clerk
1871 *Theory of Heat.* London: Longmans, Green, Chapter XXII.
1883 *Theory of Heat*, 7th ed. London: Longmans, Green.

Milne, E. A.
1952 *Sir James Jeans, A Biography.* Cambridge: Cambridge University Press. Includes comparison of views of Tyndall, Jeans, and Milne on heat death.

Momigliano, A. D.
1966 "Time in ancient historiography," *History and Theory*,

Beiheft 6: 1–23.

Nietzsche, Friedrich

1926 "Die Ewige Wiederkunft." In *Der Wille zur Macht*, reprinted in his *Gesammelte Werke*, Vol. 19. Munich: Musarion, §1053–67.

1964 *Nietzsche, An Anthology of his Works*. Edited by O. Manthey-Zorn. New York: Washington Square Press.

Penrose, L. S.

1932 "Freud's theory of Instinct and other Psycho-Biological Theories," *International Journal of Psychoanalysis* 12: 87–97. Use of energy concepts; Second Law and Fechner's stability principle.

Peterson, H.

1932 *Huxley, Prophet of Science*. London: Longmans, Green.

Pfeffer, Rose

1965 "Eternal recurrence in Nietzsche's philosophy," *Review of Metaphysics* 19: 276–300. Claims that the recurrence involves quanta of energy, not states of material systems.

Planck, Max

1897 "Ueber irreversible Strahlungsvorgänge." *Sitzungsberichte der Königlich Preussischen Akademie der Wissenschaften, Physikalisch-mathematische Klasse*, Berlin, 57–68, 715–17, 1122–45 (1897); 449–76 (1898); 440–80 (1899); *Annalen der Physik*, series 4, 1: 69–122 (1900.

Poincaré, Henri

1890 "Sur le problème des trois corps et les équations de dynamique," *Acta Mathematica* 13: 1–270.

1893 "Le mécanisme et l'expérience," *Revue de Métaphysique et de Morale* 1: 534–37. English translation in Brush (1966).

Rankine, W. J. M.

1852 "On the reconcentration of the Mechanical Energy of the Universe," *Philosophical Magazine*, series 4, 4: 358–60.

Reichenbach, Hans

1956 *The Direction of Time*. Berkeley: University of California Press.

Rey, Abel

1927 *Le Retour Éternel et la Philosophie de la Physique*. Paris: Flammarion.

Schmidt, Helmut
  1966 "Model of an oscillating cosmos which rejuvenates during contraction," *Journal of Mathematical Physics* 7: 494-509.
Schrödinger, Erwin
  1950 "Irreversibility," *Proceedings of the Royal Irish Academy* 53 A: 189-95.
Schulman, L. S.
  1973 "Correlating arrows of time," *Physical Review*, series 3D, 7: 2868-74.
Shelley, Percy Bysshe
  1822 *Hellas: A Lyrical Drama*. London: Ollier.
  1965 *The Complete Works of Percy Bysshe Shelley*, edited by Roger Ingpen and Walter Peck. Volume III. New York: Gordian Press.
Smyth, William
  1872 "Mr. Spencer and the Dissipation of Energy," *Nature* 5: 322.
Sorokin, P. A.
  1937 *Social and Cultural Dynamics*. New York: American Book Co., Vol. II, Chapter 10, "Fluctuation of the linear, cyclical and mixed conceptions of the cosmic, biological and socio-cultural processes." Correlation of atomism and materialistic or "sensate" culture.
  1947 *Society, Culture, and Personality*. New York: Harper, 676ff. Cyclic vs. linear theories of history in various cultures.
Spencer, Herbert
  1862 *First Principles*. London: Williams & Norgate. 2d ed. 1867.
  1958 *First Principles*. 4th ed. New York: DeWitt Revolving Fund.
Stambaugh, Joan
  1972 *Nietzsche's Thought of Eternal Return*. Baltimore: Johns Hopkins Press.
[Stewart, Balfour, and Tait, P. G.]
  1875 *The Unseen Universe; or, Physical Speculations on a Future State*. London: Macmillan.
Sypher, Wylie
  1962 *Loss of the Self in Modern Literature and Art*. New York: Random House, Chapter 4: "Existence and Entropy."

Szilard, L.
1929 "Über die Entropieminderung in einem thermodynamischen System bei Eingriffen intelligenter Wesen," *Zeitschrift für Physik* 53: 840–56.

Tarde, Gabriel
1890 *Les Lois de l'Imitation*. Paris: Alcan.

Terletskii, Ya. P.
1952 "O 'Fluktuatsionnoi Gipoteze' Bol'tsmana," *Zhurnal Eksperimentalnoi i Teoreticheskoi Fiziki* 22: 506–507.

Tolman, Richard C.
1934 *Relativity, Thermodynamics, and Cosmology*. Oxford: Clarendon Press.

Tyndall, John
1863 *Heat Considered as a Mode of Motion*. London: Longmans, Green. (Later editions omit "Considered" from title)
1892 "The Sabbath" (1880 lecture). In *New Fragments*. New York: Appleton, pp. 1–46. The result of tendency toward equilibrium" is not "peace and blessedness to the human race" but death.

Vogt, J. G.
1878 *Die Kraft, eine real-monistische Weltanschauung*. Leipzig: Haupt & Tischler.

Von Weizsäcker, C. F.
1939 "Der zweite Haputsatz und der Unterschied von Vergangenheit und Zukunft," *Annalen der Physik*, series 5, 36: 275–83.

Waerden, B. L. van der
1952 "Das grosse Jahr und die ewige Widerkehr," *Hermes* 80: 129–55.

White, Lynn
1942 "Christian myth and Christian history," *Journal of the History of Ideas* 3: 145–58.

Zanstra, Herman
1968 "Thermodynamics, statistical mechanics and the universe," *Vistas in Astronomy* 10: 23–43.

Zawirski, Z.
1936 *L'Evolution de la Notion du Temps*. Cracovie, Poland: Gabethner & Wolff.

Zermelo, Ernst

1896 "Über einen Satz der Dynamik und die mechanishe Wärmetheorie," *Annalen der Physik,* series 3, 57: 485–94. English translation in Brush (1966).
1896 "Ueber mechanische Erklärungen irreversibler Vorgänge," *Annalen der Physik,* series 3, 59: 793–801. English translation in Brush (1966).

Chapter VI

Adams, Henry
  1918 *The Education of Henry Adams.* Boston: Massachusetts Historical Society, Chapter XXXI.
Aliotta, Antonio
  1912 *La Reazione Idealistica contro la Scienza.* Palermo: Casa Editrice "Optima." English translation, *The Idealistic Reaction Against Science.* London: Macmillan, 1914.
Arnheim, Rudolf
  1971 *Entropy and Art. An Essay on Disorder and Order.* Berkeley: University of California Press.
Artigiani, Philip R.
  1969 *The Functional Self: A Study of the Effects of the Philosophy and Practice of Science on the Scientist's Self-Image During the Nineteenth and Twentieth Centuries.* Ph.D. Dissertation, American University.
Avenarius, Richard
  1876 *Philosophie als Denken der Welt gemäss dem Prinzip des kleinstein Kraft-masses.* Leipzig: Fues.
  1888 *Kritik der reinen Erfahrung.* Leipzig: Fues, 1888–90.
Ayres, C. E.
  1932 *Huxley.* New York: Norton. (See pp. 118ff. for critique of positivism.)
Badash, Lawrence
  1972 "The Completeness of Nineteenth Century Science," *Isis* 63: 48–58.
Barker, Ernest
  1915 *Political Thought in England from Herbert Spencer to the Present Day.* New York: Holt.
Benda, Julien

1927 *La Trahison des Clercs*. Paris: Grasset.
Bergson, Henri
1889 *Essai sur les données immédiates de la conscience*. Paris: Alcan. English translation: *Time and Free Will*. New York: Macmillan, 1910.
Berthelot, René
1911 *Un Romantisme Utilitaire. Étude sur le Mouvement Pragmatiste*. Paris: Alcan. Nietzsche and Poincaré.
Biermann, Kurt-R.
1973 "Kronecker, Leopold," *Dictionary of Scientific Biography*, edited by C. C. Gillispie, vol. 7, pp. 505–9. New York: Scribner.
Boltzmann, Ludwig
1872 "Weitere Studien über das Warmegleichgewicht unter Gasmolekülen," *Sitzungsberichte der kaiserlichen Akademie der Wissenschaften in Wien*, Abt. 2, 66: 275–370.
1886 "Der zweite Haputsatz der mechanische Wärmetheorie," *Almanach der kaiserlich Akademie der Wissenschaften, Wien* 36: 225–59.
1964 *Lectures on Gas Theory*. Translated by Stephen G. Brush. Berkeley: University of California Press.
Boring, E. G.
1942 "Human nature vs. Sensation: William James and the Psychology of the Present," *American Journal of Psychology* 55: 310–27. Dichotomy of phenomenology vs. reduction.
1950 *A History of Experimental Psychology*, 2d ed. New York: Appleton-Century-Crofts, Chapters 14–20.
Boutroux, Émile
1874 *De la Contingence des Lois de la Nature*. Paris: Bailliere.
Brown, Alan Willard
1947 *The Metaphysical Society: Victorian Minds in Crisis, 1869–1880*. New York: Columbia University Press.
Brush, Stephen G.
1969 "Romance in Six Figures," *Physics Today* 22 (1) (January): 9.
Buchanan, James
1857 *Modern Atheism under its forms of Pantheism, Materialism, Secularism, Development, and Natural Laws*. Boston: Gould and Lincoln.

Büttner, Alexander
   1911 *Von der Materie zum Idealismus: Skizze eines einheitlichen Weltbildes.* Crefeld: Furst/Schäckermann & de Greiff.
Cameron, F. K.
   1900 "Some objections to the atomic theory," *Science* 11: 608–12.
Carbonnelle, I.
   1877 "L'Aveuglement Scientifique," *Revue des Questions Scientifique* 1: 5–53, 512–61, 2: 236–73 (1877); 3: 548–88, 4: 578–624 (1878); 5: 234–86, 6: 196–232 (1879). Religious and philosophical implications of science, atomism, materialism, thermodynamics.
Caro, Elme Marie
   1867 *Le Matérialisme et la Science.* Paris: Hachette.
Carpenter, Edward
   1889 *Civilisation; its Cause and Cure.* London: Sonnenschein.
Carus, Paul
   1890 "The reaction against materialism," *Open Court* 4: 2169–72.
Cassirer, E.
   1936 "Determinismus und Indeterminismus in der modernen Physik," *Göteborgs Högskolas Årsskrift*, 42, part 3. English translation, *Determinism and Indeterminism in Modern Physics.* New Haven: Yale University Press, 1956, Part 1.
Clark, Peter
   See Howson
Clark, Xenos
   1888 "Free Will a Mechanical Possibility," *Open Court* 2: 975–77.
Comte, Auguste
   1830 *Cours de Philosophie Positive.* Paris: Bachelier, 1830–42.
Cournot, Antoine Augustin
   1851 *Essai sur les fondements de nos connaissances et sur les caractères de la critique philosophique.* Paris: Hachette.
Cravens, Hamilton
   1971 "The Abandonment of Evolutionary Social Theory in America: The Impact of Academic Professionalization upon American Sociological Theory, 1890–1920," *American Studies* 12: 5–20.
Croll, James
   1872 "What determines molecular motion?—The Fundamental

Problem of Nature," *Philosophical Magazine*, series 4, 44: 1–25.
1891 *The Philosophical Basis of Evolution*. London: Stanford.

Dingle, Herbert
1951 "Philosophy of Physics 1850–1950," *Nature* 168: 630–36.

Drake, Stillman
1959 "J. B. Stallo and the Critique of Classical Physics." In *Men and Moments in the History of Science*, edited by H. M. Evans, pp. 22–37. Seattle: University of Washington Press.

Du Bois-Reymond, Emil
1874 "Ueber die Grenzen des naturwissenschaftlichen Erkenntnis." *Tageblatt der 1872 Versammlung Deutscher Naturforscher und Aerzte*, pp. 85–86. English translation in *Popular Science Monthly* 5: 17–32 (1874).

Duhem, Pierre
1906 *La Theorie Physique, son Objet et sa Structure*. Paris: Chevalier et Riviera. English translation, *The Aim and Structure of Physical Theory*. Princeton, N.J.: Princeton University Press, 1954.

Eisen, Sydney
1964 "Huxley and the positivists," *Victorian Studies* 7: 337–58.

Elkana, Yehuda
1974 "Boltzmann's scientific research programme and its alternatives." In *The Interaction between Science and Philosophy*, edited by Y. Elkana, pp. 243–279. Atlantic Highlands, N. J.: Humanities Press.

Engels, Friedrich
1940 *Dialectics of Nature*. New York: International Publishers.

Enriques, Frédéric
1909 *Les Problèmes de la Science et la Logique*. Paris: Alcan.
1913 *Les Concepts Fondamentaux de la Science: Leur Signification réelle et leur Acquisition Psychologique*. Paris: Flammarion.

Eve, A. S., and Creasey, C. H.
1945 *Life and Work of John Tyndall*. London: Macmillan, Chapter XV, The Belfast Address.

Feuillerat, Albert
1937 *Paul Bourget: Histoire d'un Esprit sous la Troisième République*. Paris: Plon.

FitzGerald, G. F.
  1896 "Ostwald's Energetics," *Nature* 53: 441.
Flugel, Otto
  1865 *Des Materialismus vom Standpunkte der atomistisch-mechanischen Naturforschung beleuchtet.* Leipzig: Pernitzsch.
Fouillée, A. J. E.
  1896 *Le Mouvement Idéaliste et la Réaction contre la Science Positive.* Paris: Alcan.
Frank, Philipp
  1937 "The mechanical versus the mathematical conception of nature," *Philosophy of Science* 4: 41–74.
  1941 *Modern Science and its Philosophy.* Cambridge, Mass.: Harvard University Press.
Freeman, Derek
  1966 "Social anthropology and the scientific study of human behavior," *Man*, new series, 1: 330–40.
Galton, Francis
  1872 "Statistical inquiries into the efficacy of prayer," *Fortnightly Review* (n.s.) 12: 125–35.
Garland, Hamlin
  1894 *Crumbling Idols: Twelve Essays Dealing Chiefly with Literature, Painting and the Drama.* Chicago and Cambridge, Mass.: Stone and Kimball. Reprinted with a new introduction by Jane Johnson, Cambridge, Mass.: Belknap Press of Harvard University Press, 1960. A manifesto for "veritism" (a version of realism) and against romanticism.
Gasman, Daniel
  1970 *The Scientific Origins of National Socialism, Social Darwinism in Ernst Haeckel and the German Monist League.* London: Macdonald; New York: American Elsevier.
Glass, Bentley
  1953 "The long neglect of a scientific discovery: Mendel's laws of inheritance." In *Studies in Intellectual History* by G. Boas *et al.* pp. 148–60. Baltimore: The Johns Hopkins Press.
Guerlac, H. E.
  1951 "Science and French National Strength." In *Modern France*, edited by E.M. Earle, pp. 81–105. Princeton: Princeton University Press.

Griffiths, Richard
  1966 *The Reactionary Revolution: The Catholic revival in French literature, 1870–1914*. London: Constable.
Hahn, Roger
  1965 "Laplace's first Formulation of Scientific Determinism in 1773," *Actes du XIe Congrès International d'Histoire des Sciences, Cracow, 1965,* 2: 167–71 (1968).
Haines, George, IV
  1969 *Essays on German Influence upon English Education and Science 1850–1919*. Connecticut College Monograph No. 9. Hamden, Conn.: Archon Books.
Hayes, C. J. H.
  1941 *A Generation of Materialism, 1871–1900*. New York: Harper & Row.
Hermann, Armin, and Kaiser, Walter
  1972 "Der Positivismus in der Physik des 18. und 19. Jahrhunderts," *Rete* 1: 135–44.
Hibben, John Grier
  1903 "The Theory of Energetics and its Philosophical Bearings," *Monist* 13: 321–30. Critique of Ostwald.
Hiebert, Erwin N.
  1971 "The energetics controversy and the new thermodynamics." In *Perspectives in the History of Science and Technology*, edited by D.H.D. Roller, pp. 67–86. Norman: University of Oklahoma Press.
Howson, Colin, ed.
  1976 *Method and Appraisal in the Physical Sciences*. New York: Cambridge University Press. Includes: Imre Lakatos, "History of Science and its Rational Reconstructions," pp. 1–39; Peter Clark, "Atomism versus Thermodynamics," pp. 41–105.
Hoyle, Fred
  1956 *Men and Materialism*. New York: Harper.
Hughes, H. Stuart
  1958 *Consciousness and Society: The Reconstruction of European Social Thought, 1890–1930*. New York: Knopf.
Huxley, T. H.
  1868 "On the Physical Basis of Life." In *Collected Essays*. Vol. 1,

pp. 130–65. New York: Macmillan, 1893 (Lecture in Edinburgh, 1868).
1871 *Lay Sermons, Addresses, and Reviews.* New York: Appleton.
1948 *Selections from the Essays of T. H. Huxley,* edited by Alburey Castell. New York: Appleton-Century-Crofts.

Jaki, Stanley L.
1966 *The Relevance of Physics.* Chicago: University of Chicago Press.

Jammer, Max
1961 *Concepts of Mass in Classical and Modern Physics.* Cambridge, Mass.: Harvard University Press, Chapter 8.

Jellett, J. H.
1874 Presidential Address to the Mathematics and Physics Sections of the British Association meeting at Belfast, *Nature* 10: 319–24.

Jensen, J. Vernon
1970 "The X-Club: Fraternity of Victorian scientists," *British Journal for the History of Science* 5: 63–72.

Jodl, Friedrich
1891 "German Philosophy in the Nineteenth Century," *Monist* 1: 263–77.

Joyce, C. R. B., and Welldon, R. M. C.
1965 "The Objective Efficacy of Prayer: A Double-Blind Clinical Trial," *Journal of Chronic Diseases* 18: 367–77.

[Kelvin] Thomson, William
1870 "The Size of Atoms," *Nature* 1: 551–53.

Klein, Martin J.
1972 "Mechanical explanation at the End of the Nineteenth Century," *Centaurus* 17: 58–82.

Kleinpeter, H.
1905 *Die Erkenntnistheorie der Naturforschung der Gegenwart, Unter Zugrundelegung der Anschauungen von Mach, Stallo, Clifford, Kirchhoff, Hertz, Pearson und Ostwald.* Leipzig: Barth.
1913 *Der Phänomenalismus, eine naturwissenschaftliche Weltanschauung.* Leipzig: Barth.

Kuhn, Wolfgang
1964 "Auschwitz—Ende einer 'Biologischen Weltanschauung,'" *Stimmen der Zeit* 174 (7): 36–49.

Lakatos, Imre
  See Howson
Lange, F. A.
  1866 *Geschichte des Materialismus und Kritik seiner Bedeutung in der Gegenwart.* Iserlohn: Baedeker. English translation, *History of Materialism and Criticism of its Present Importance.* London: Trübner, 1879–81.
Laplace, P. S. de
  1773 "Recherches sur l'integration des equations differentielles aux differences finies, et sur leur usage dans la theorie des hasards," *Memoires de Mathematique et de Physique presentés a l'Academie Royale des Sciences par Divers Savans* 7: 37–163. (Quoted by Hahn 1965.)
  1814 *Essai Philosophique sur les Probabilites.* Paris: M$_{me}$ Ve Courcier. English translation: *A Philosophical Essay on Probabilities.* London: Chapman & Hall, 1902.
Lenin, V. I.
  1909 *Materializm i Empiriokrititsism; Kriticheskie Zametki ob odnoi reaktsionnoi Filosofii.* Moscow: Izdanie "Zveno"; 2d ed. 1920. English translation, *Materialism and Empirio-Criticism: Critical Comments on a Reactionary Philosophy.* Moscow: Cooperative Pub. Soc. of Foreign Workers in the USSR, 1937. Reprinted by Foreign Languages Pub. House, Moscow, 1947.
Lindbergh, Charles
  1948 *Of Flight and Life.* New York: Scribners.
Littledale, Richard Frederick
  1872 "The Rationale of Prayer," *Contemporary Review* 20: 430–54.
Littré, E.
  1864 "Preface d'un Disciple," in Auguste Comte's *Cours de Philosophie Positive* (1830).
Lodge, Oliver
  1891 "Force and determinism," *Nature* 43: 491; 44: 198, 272–73. Other notes on this subject in *Nature* by Morgan, Dixon, Wetterham, and Sherlock.
Loschmidt, Josef
  1865 "Zur Grosse der Luftmolecule," *Sitzungsberichte der kaiserlichen Akademie der Wissenschaften in Wien,* Abt. 2, 52: 395–413.

Mach, Ernst
   1872 *Die Geschichte und die Wurzel des Satzes von der Erhaltung der Arbeit.* Prague: Calve. English translation: *History and Root of the Principle of the Conservation of Energy.* Chicago: Open Court, 1911.
   1882 "Die ökonomische Natur der physikalischen Forschung." *Almanach der kaiserlichen Akademie der Wissenschaften, Wien* 32: 293–319. English translation in *Popular Scientific Lectures.*
   1883 *Die Mechanik in ihrer Entwickelung historisch-kritisch dargestellt.* Leipzig: Brockhaus. English translation, *The Science of Mechanics, A Critical and Historical Exposition of its Principles.* Chicago: Open Court, 1893.
   1886 *Beiträge zur Analyse der Empfindungen.* Jena: Fischer. English translation, *Contributions to the Analysis of the Sensations.* Chicago: Open Court, 1897.
   1895 *Popular Scientific Lectures.* Chicago: Open Court.
   1896 *Die Principien der Wärmelehre. Historisch-kritisch entwickelt.* Leipzig: Barth.
   1905 *Erkenntnis und Irrtum. Skizzen zur Psychologie der Forschung.* Leipzig: Barth, 1905.
MacLeod, Roy M.
   1969 "The X-Club: A Social network of science in late-Victorian England," *Notes and Records of the Royal Society of London* 24: 305–22.
Matson, F. W.
   1964 *The Broken Image: Man, Science, and Society.* New York: Braziller.
M'Cosh, James
   1872 "On Prayer," *Contemporary Review* 20: 777–82.
Means, John O., ed.
   1876 *The Prayer-Gauge Debate.* Boston: Congregational Publishing Society.
Meyer, D. H.
   1962 "Paul Carus and the Religion of Science," *American Quarterly* 14: 597–607.
Michelson, A. A.
   1902 *Light Waves and Their Uses.* Chicago: University of Chicago Press. Lectures at the Lowell Institute, Boston, 1899.

Millikan, Robert A.
  1927 "Conceptions in physics changed in our generation," *Scientia* 41: 255–64.
  1950 *The Autobiography of Robert A. Millikan.* New York: Prentice-Hall.
Mosse, George L.
  1964 *The Crisis of German Ideology: Intellectual Origins of the Third Reich.* New York: Grosset & Dunlap.
Mott, N. F.
  1949 "Physical Science and the Beliefs of the Victorians." In *Ideas and Beliefs of the Victorians*, pp. 215–21. London: Sylvan Press.
Nye, Mary Jo
  1974 "Gustave LeBon's Black Light: A Study in Physics and Philosophy in France at the Turn of the Century," *Historical Studies in The Physical Sciences* 4: 163–95. A "discovery" influenced by neoromanticism.
Packard, Vance
  1960 *The Waste Makers.* New York: McKay.
Parrington, Vernon Louis
  1930 *The Beginnings of Critical Realism in America: 1860–1920.* New York: Harcourt, Brace and World.
Parsons, Talcott
  1937 *The Structure of Social Action. A Study in Social Theory with special reference to a group of recent European writers.* New York: McGraw-Hill.
Paul, Harry W.
  1968 "The debate over the bankruptcy of science in 1895," *French Historical Studies* 5: 299–327.
  1971 "Science and the Catholic Institutes in Nineteenth-Century France," *Societes, A Review of Social History* 1: 271–85.
  1972 "The Crucible and the Crucifix: Catholic Scientists in the Third Republic," *Catholic Historical Review* 58: 195–219.
  1972 "The issue of decline in nineteenth-century French science," *French Historical Studies* 7: 416–50.
Pearson, Karl
  1892 *The Grammar of Science.* London: Block.
Peirce, C. S.
  1892 "The Doctrine of Necessity examined," *Monist* 2: 321–37.

1958 *Values in a Universe of Chance*, edited by Philip P. Wiener. Garden City, N.Y.: Doubleday Anchor.

Peterson, Houston

1932 *Huxley, Prophet of Science*. London: Longmans, Green.

Phillips, D. C.

1970 "Organicism in the late 19th and early 20th centuries," *Journal of the History of Ideas* 31: 413–32.

Poincaré, Henri

1902 *La Science et l'Hypothesis*. Paris: Flammarion. English translation, *Science and Hypothesis*. New York: Science Press, 1905.

1905 *La Valeur de la Science*. Paris: Flammarion. English translation, *The Value of Science*. New York: Science Press, 1907.

1908 *Science et Méthode*. Paris: Flammarion. English translation, *Science and Method*. London: Nelson, 1914.

Primer, Sylvester

1908 "The Influence of Science upon German Literature, based on Haeckel's Weltraethsel and Nietzsche's Philosophie," *Transactions of the Texas Academy of Science* 11: 54–68 (1908–1909).

Rey, A.

1907 *La Théorie de la Physique chez les Physiciens Contemporains*. Paris: Alcan.

Ringer, Fritz K.

1969 *The Decline of the German Mandarins. The German Academic Community, 1890–1933*. Cambridge, Mass.: Harvard University Press.

Romanell, Patrick

1956 "Romanticism and Croce's conception of Science," *Review of Metaphysics* 9: 505–14.

Roosevelt, Theodore

1960 "The greater goal," *This Week*, Dec. 18, p. 2. Reprinted from *Metropolitan*, November 1918.

Sageret, J.

1920 *La Vague Mystique*. Paris: Flammarion.

Schorske, C. E.

1961 "Politics and the Psyche in *fin-de-siècle* Vienna: Schnitzler and Hofmannsthal," *American Historical Review* 66: 930–46.

Schuster, Arthur

1918 *Britain's Heritage of Science.* London: Constable.

Scott, Wilson L.
1970 *The Conflict Between Atomism and Conservation Theory, 1644 to 1860.* New York: Elsevier.

Simon, W. M.
1963 *European Positivism in the Nineteenth Century; An Essay in Intellectual History.* Ithaca: Cornell University Press. States that the first public notice of Comte's philosophy in England was a review by the physicist David Brewster in the *Contemporary Review*, 1838.

Sorokin, P. A.
1937 *Social and Cultural Dynamics.* New York: American Book Co. Chapter 4, Fluctuation of Idealism and Materialism. See also Appendix.

Stallo, J. B.
1960 *The Concepts and Theories of Modern Physics.* New York: Appleton. Reprinted with a new introduction by Percy W. Bridgman, Cambridge, Mass.: Harvard University Press, 1960.

Stebbing, Susan
1958 *Philosophy and the Physicists.* New York: Dover.

Stern, Fritz
1961 *The Politics of Cultural Despair.* Berkeley: University of California Press, Chapter 8, "Art and the Revolt Against Modernity."

Thiele, Joachim
1968 " 'Naturphilosophie' und 'Monismus' um 1900," *Philosophia Naturalis* 10: 295–315.

[Thompson, Henry]
1872 "The 'Prayer for the Sick': Hints toward a serious attempt to estimate its value," *Contemporary Review* 20: 206–10.

Thomson, David
1962 "Social and Political Thought." In *The New Cambridge Modern History*, Vol. XI, *Material Progress and World-Wide Problems, 1870–1898.* Cambridge: Cambridge University Press.

Tuchman, Barbara
1966 *The Proud Tower: A Portrait of the World Before the War, 1890–1914,* New York: Macmillan.

Turner, Frank
  1974a "Rainfall, Plagues, and the Prince of Wales: A Chapter in the Conflict of Religion and Science," *Journal of British Studies* 13: 46–65.
  1974b *Between Science and Religion. The Reaction to Scientific Naturalism in Late Victorian England.* New Haven: Yale University Press.
Tyndall, John
  1868 Address to the Mathematical and Physical Section of the British Association. *Report of the 38th Meeting of the British Association for the Advancement of Science,* 1–6.
  1872a Introduction to "The 'Prayer for the Sick' " [by Henry Thompson]. *Contemporary Review* 20: 205–206.
  1872b "On Prayer," *Contemporary Review* 20: 763–66.
  1897 *Fragments of Science.* Volume 2. New York: Appleton.
Tyndall, John; Galton, Francis; et al.
  1876 *The Prayer-Gauge Debate,* edited by J. O. Means. Boston: Congregational Publishing Society.
Virtanen, Reino
  1965 "Marcelin Berthelot: A Study of a Scientist's Public Role." *University of Nebraska Studies, No. 31.*
Waals, Johannes Diderik van der
  1873 *Over de Continuiteit van den Gas- en Vloeistoftoestand.* Dissertation, Leiden.
Ward, James
  1899 *Naturalism and Agnosticism.* New York: Macmillan.
Welsh, Alexander
  1973 "Theories of Science and Romance, 1870–1920," *Victorian Studies* 17, 135–54.
Whyte, L. L.
  1965 "Atomism, structure and form. A report on the Natural Philosophy of Form." In *Structure in Art and Science,* ed. Gyorgy Kepes. New York: Braziller, pp. 20–28.
Wilde, Oscar
  1972 "Lecture on the English Renaissance" (1906). In his *Works,* Vol. 9, pp. 103–17. New York: AMS Press.
Wilson, James M.
  1919 "Science and the Church," *Nature* 104: 201–202. Recalls the

antireligious attitudes of scientists 50 years earlier, especially Tyndall and his school.

Young, John
1871 "From Geology to History," *Transactions of the Geological Society of Glasgow* 3: 341–67.

Zoellner, J.
1872 *Über die Natur der Cometen. Beiträge zur Geschichte und Theorie der Erkenntniss*, 2d ed. Leipzig: Staackmann.

Chapter VII

Allen, Garland
1976 "Genetics, Eugenics and Society: Internalists and Externalists in Contemporary History of Science," *Social Studies of Science* 6: 105–22.

Andersson, Ola
1962 *Studies in the Prehistory of Psychoanalysis: The Etiology of Psychoneuroses and some related themes in Sigmund Freud's Scientific Writings and Letters*. Studia Scientiae Paedagogicae Upsaliensis, 3. Stockholm: Svenska Bokförlaget.

Balaguer Periguell, Emilio
1969 "El somaticismo y la doctrina de la 'degeneración' en la psiquiatría valenciana del siglo XIX," *Medicina Española* 62: 388–94.

Barzun, Jacques
1962 "From the Nineteenth Century to the Twentieth." In *Chapters in Western Civilization*, 3rd ed., ed. Bernard Wishy. New York: Columbia University Press, Vol. II, pp. 340–63.

Baudelaire, Charles
1857 *Les Fleurs du Mal*. Paris: Poulet-Malassis et de Broise.
1946 "Une Charogne," English translation by G. Wagner, in Baudelaire's *Selected Poems*. London: Falcon, p. 49.

Beale, Octavius Charles
1911 *Racial Decay: A Compilation of Evidence from World Sources*. London: Fifield.

Beard, George M.
1881 *American Nervousness. Its Causes and Consequences*. New York: Putnam.

1883 *Herbert Spencer on American Nervousness. A Scientific Coincidence.* New York: Putnam.
Bitterlich, Max
   1932 *Die Entartung des Menschen, das Negativ seiner Veredlung. Ein Naturgesetz.* Wien: Gerold.
Boulenger, M., and Ensch, N.
   1905 *Hygiène scolaire. La Lutte contre la dégénérescence en Angleterre.* Bruxelles: Misch & Thron.
Brand, Lilian
   1910 "Alcoholism and social problems," *The Survey* 25 (October): 17–21. Attempts to explain away the Elderton-Pearson results.
Bunke, Oswald
   1911 *Kultur und Entartung.* Berlin: Springer, 2. Aufl. 1922.
Carrere, Jean
   1921 *Les Mauvais Maitres.* Paris: Plon. English translation, *Degeneration in the Great French Masters.* New York: Brentano, 1922.
Carter, A. E.
   1958 *The Idea of Decadence in French Literature, 1830–1900.* Toronto: Toronto University Press.
Chase, Allan
   1975 "Eugenics vs Poor White Trash: The Great Pellagra Cover-up," *Psychology Today* 8 (9) (February): 82–86.
Clifford, W. K.
   1876 Letter to F. Pollock, July 15, 1876. In *Lectures and Essays,* Vol. 1, pp. 58–59. London: Macmillan.
Copland, Aaron
   1968 *The New Music 1900–1960.* New York: Norton.
Cowan, Ruth Schwartz
   1968 "Sir Francis Galton and the continuity of germ-plasm: A biological idea with political roots," *Actes du XIIe Congrès Internationale d'Histoire des Sciences, 1968,* 8: 181–86. Paris: Hermann (1971).
   1972 "Francis Galton's contribution to genetics," *Journal of the History of Biology* 5: 389–412.
Crafts, Wilbur F.
   1918 *Why Dry? Briefs for Prohibition . . .* Washington, D.C.: In-

ternational Reform Bureau (1919).

Damm, Alfred
1895 *Die Entartung der Menshen und die Beseitigung der Entartung.* Berlin: Rouschel.

Davenport, Charles B.
1911 *Heredity in Relation to Eugenics.* New York: Holt. Reprinted, with new introduction by C. E. Rosenberg, New York: Arno, 1972.

Doran, R. E.
1903 "A consideration of the hereditary factors in epilepsy," *American Journal of Insanity* 60: 61–73.

Elderton, Ethel M., and Pearson, Karl
1910 *A First Study of the Influence of Parental Alcoholism on the Physique and Ability of the Offspring.* London: Dulau.

Elwood, Everett S.
1914 "Mental Defect in Relation to Alcohol with some Notes on Colonies for Alcoholic Offenders," *Proceedings of National Conference of Charities and Correction, 41st Session,* pp. 306–14. Fort Wayne Ind.: Fort Wayne Printing Co. Evidence for and against hereditary effects of alcohol.

Fehlinger, Hans
1919 *Rassenhygiene. Beiträge zur Entartungsfrage.* Langensalza: Wandt & Klauwell.

Féré, C. S.
1888 *Dégénérescence et criminalité, Essai Physiologique.* Paris: Alcan.

Ferguson, Donald N.
1935 *A History of Musical Thought.* New York: Crofts.

Fink, Arthur E.
1938 *Causes of Crime. Biological Theories in the United States 1880–1915.* Philadelphia: University of Pennsylvania Press.

Fischer-Homberger, Esther
1971 "Charcot und die Ätiologie der Neurosen," *Gesnerus* 28: 35–46.

Foster, Milton P.
1954 *The Reception of Max Nordau's Degeneration in England and America.* Ph.D. Dissertation, University of Michigan.

Freud, Sigmund
1893 "Heredity and the Aetiology of the Neuroses." In *The Stan-*

dard Edition of the Complete Psychological Works of Sigmund Freud, Vol. III (1893–1899), Early Psychoanalytic Publications, pp. 143–56. London: Hogarth, 1962.

Friedlander, Ruth
 1973 *Benedict-Augustin Morel and the development of the theory of degenerescence.* Ph.D. Dissertation, University of California, San Francisco.

Galippe, Victor
 1905 *L'Hérédité des Stigmates de Dégénérescence et les Familles Souveraines.* Paris: Masson. Pictorial and descriptive.

Galton, Francis
 1869 *Hereditary Genius, An Inquiry into its Laws and Consequences.* London: Macmillan.

Gasman, Daniel
 1971 *The Scientific Origins of National Socialism. Social Darwinism in Ernst Haeckel and the German Monist League.* London: Macdonald; New York: American Elsevier.

Genil-Perrin, G.-P.-H.
 1913 *Histoire des Origines et de l'Evolution de l'Idée de Dégénérescence en Medicine Mentale.* Paris: Faculté de Medecine.

Gold, Milton
 1960 "The early psychiatrists on degeneracy and genius," *Psychoanalysis and the Psychoanalytic Review* 47: 37–55 (1960–61).
 1961 "The continuing 'degeneration controversy,' " *Bucknell Review* 10: 87–101.

Grant, Madison
 1917 *The Passing of the Great Race.* London: Bell.

Gustafson, Axel
 1887 *The Foundation of Death: A Study of the Drink-Question.* 3rd ed. Boston: Heath, Chapter VIII.

Haller, John S., Jr.
 1971a *Outcasts from Evolution: Scientific Attitudes of Racial Inferiority, 1859–1900.* Urbana: University of Illinois Press.
 1971b "Neurasthenia: The Medical Profession and the 'New Woman' of the late 19th century," *New York State Journal of Medicine* 71: 473–82.

Haller, Mark H.
　1963 *Eugenics. Hereditarian Attitudes in American Thought.* New Brunswick, N. J.: Rutgers University Press.
Hellpach, Willy
　1902 *Nervosität und Kultur.* Berlin: Raede.
Hildebrandt, Kurt
　1939 *Norm, Entartung, Verfall: Bezogen auf den Einzelnen, die Rasse, den Staat.* Stuttgart: Kohlhammer.
Hirsch, William
　1894 *Genie und Entartung, Eine psychologische Studie.* Berlin and Leipzig: Coblentz. English translation, *Genius and Degeneration, a psychological Study.* New York: Appleton, 1896.
Hobson, Richmond Pearson
　1917 "Destroying the Great Destroyer," *Congressional Record* 55: 7820–26.
Holmes, Samuel J.
　1921 *The Trend of the Race: A Study of Present Tendencies in the Biological Development of Civilized Mankind.* New York: Harcourt, Brace & Co., Chapter XII.
　1924 *A Bibliography of Eugenics,* University of California Publications in Zoology, Volume 25. Berkeley: University of California Press. "The problem of degeneracy," pp. 81–101; "Alcoholism in relation to heredity; Lead poisoning; Blastophthoria," pp. 185–209.
Horsley, Victor, and Sturge, Mary D.
　1907 *Alcohol and the Human Body.* London and New York: Macmillan. 2nd ed. 1908, Chapter XV.
Howe, Samuel Gridley
　1858 *On the Causes of Idiocy.* New York: Arno Press, 1972 (reprint of 1858 ed.).
Jones, Bartlett C.
　1963 "Prohibition and Eugenics 1920–1933," *Journal of the History of Medicine* 18: 158–72.
Jordan, D. S.
　1906 *The Blood of the Nation: A Study of the Decay of Races through the Survival of the Unfit.* Boston: American Unitarian Association.

1915 *War and the Breed; the Relation of War to the Downfall of Nations.* Boston: Beacon.
Josephson, Matthew
    1928 *Zola and his Time.* New York: Macaulay.
Kende, Moriz
    1901 *Die Entartung des Menschengeschlechts, ihre Ursachen und die Mittel zu ihrer Bekämpfung.* Halle: Marhold.
Kern, Stephem
    1974 "Explosive intimacy: psychodynamics of the Victorian family," *History of Childhood Quarterly* 1: 437–61.
Koren, John
    1916 *Alcohol and Society.* New York: Holt. Claims alcohol does *not* cause hereditary degeneration.
Krauss, Franz
    1903 *Der Völkertod. Eine Theorie der Dekadenz.* Leipzig: Deuticke.
Lange, Frederik
    1907 *Degeneration in Families. Observations in a Lunatic Asylum.* Translated from Danish. London: Kimpton.
Legrain, Paul-Maurice
    1889 *Hérédité et Alcoolisme: Étude Psychologique et Clinique sur les dégénérés buveurs et les familles d'ivrogenes.* Paris: Doin.
    1895 *Dégénérescence Sociale et Alcoolisme.* Paris: Carré.
Leibbrand, Werner, and Wettley, Annemarie
    1961 *Der Wahnsinn. Geschichte der Abendlandischen Psychopathologie.* München: Alber, pp. 519–45. Review of writings of Prosper Lucas, Buchez, Morel, Magnan, Charcot Krafft-Ebing, Kraepelin.
Leppmann, A. and F.
    1909 "Alcoholism and morphinism." In *Marriage and Disease,* edited by Hermann Senator and S. Kaminer, Vol. 2, pp. 1057–1133. New York: Hoeber.
Lombroso, Cesare
    1907 *Genio e degenerazione. Nuovi studi e nuovi battaglie.* 2d ed. Milano: Sandron.
Lucas, Prosper
    1847 *Traité philosophique et physiologique de l'hérédité naturelle dans les états de santé et de maladie du système nerveux.* Paris: Balliere, 1847–50.

Ludmerer, Kenneth M.
  1972 *Genetics and American Society.* Baltimore: The Johns Hopkins Press.
Lydston, G. F.
  1904 *The Diseases of Society (The vice and crime problem).* Philadelphia: Lippincott.
MacDonald, Arthur
  1898 "Emile Zola, A Psycho-Physical Study," *Open Court* 12: 467–94. Reprinted as a pamphlet under same title, Washington, D.C., 1901.
Magnan, Valentin
  1893 *Recherches sur les centres nerveux, alcoolisme, folie des héréditaires dégénérés, paralysie générale, médecine légale,* 2e série. Paris: Masson.
Magnan, Valentin, and Legrain, Paul-Maurice
  1895 *Les dégénérés (état mental et syndromes épisodiques).* Paris: Rueff.
Martindale, Colin
  1971 "Degeneration, disinhibition, and genius," *Journal of the History of Behavioral Sciences* 7: 177–82.
Martineau, Henry
  1907 *Le Roman Scientifique d'Émile Zola: La Médecine et les Rougon-Macquart.* Paris: Bailliere.
Mason, R. Osgood
  1901 "The curse of inebriety," *Arena* 26: 128–36.
Mason, S.
  1914 *Bibliography of Oscar Wilde.* London: Laurie.
McKim, W. D.
  1900 *Heredity and Human Progress.* New York: Putnam. McKim advocated eugenics and the painless extinction of idiots and habitual criminals.
Miller, E. C. L.
  1905 "Alcoholism and degeneration," *Independent* 58: 261–62. Review of Bunge's statistics.
Milner, G.
  1931 *The Problem of Decadence.* London: Williams and Norgate.
Möbius, P. J.
  1900 *Ueber Entartung.* Wiesbaden: Bergmann.

Morel, Benedict Augustin
   1857 *Traité des Dégénérescences Physiques, Intellectuelles et Morales de l'Espèce Humaine, et des Causes qui produisent ces variétés maladives.* Paris: Bailliere.
Mosse, George L.
   See Nordau, Max.
Mott, F. W.
   1905 "A discussion on the relationship of heredity to disease," *British Medical Journal* 2: 1086–91.
Nation, Carry Amelia
   1908 *The use and need of the Life of Carry A. Nation.* Rev. ed. Topeka, Kan.: Steves.
Nordau, Anna and Maxa
   1943 *Max Nordau, A Biography.* translated from French. New York: The Nordau Committee.
Nordau, Max
   1892 *Entartung.* Berlin: Duncker.
   1968 *Degeneration.* Translated from the Second Edition of the German Work. With an Introduction by George L. Mosse. New York: Fertig.
Pearson, Karl
   1910 *Supplement to the Memoir entitled: The Influence of Parental Alcoholism on the Physique and Ability of the Offspring: A Reply to the Cambridge Economists.* London: Dulau.
   1911 "Alcoholism and Degeneration," *British Medical Journal* 2: 221–29.
Pearson, Karl, and Edlerton, Ethel M.
   1910 *A Second Study of the Influence of Parental Alcoholism on the Physique and Ability of the Offspring; Being a Reply to certain medical critics of the first memoir and an examination of the rebutting evidence cited by them.* Eugenics Laboratory Memoirs, No. 13. London: Francis Galton Laboratory for National Eugenics.
Petrazzani, P.
   1911 *Le Degenerazioni Umane (Studio di Biologia Clinia).* Milano: Vallardi.
Pickens, Donald K.
   1968 *Eugenics and the Progressives.* Nashville: Vanderbilt University Press.

Pickett, Deets, ed.
  1917 *The Cyclopedia of Temperance, Prohibition, and Public Morals.* New York and Cincinnati: The Methodist Book Concern, pp. 186–91.
Popenoe, Paul, and Johnson, Roswell H.
  1918 *Applied Eugenics.* New York: Macmillan.
Potts, W. A.
  1905 "Causation of mental defect in children," *British Medical Journal* 2: 946–48.
Raven, C. E.
  1960 "The Impact of Physics on Science and Religion." In *A Physics Anthology*, edited by N. Clarke, pp. 33–46. London: Chapman & Hall.
Reid, G. Archdall
  1902 *Alcoholism: A Study in Heredity.* London: Bailliere, Tindall & Cox.
Rentoul, R. R.
  1906 *Race Culture; or, Race Suicide? (A Plea for the Unborn).* London: Scott.
Rosanoff, M. A. and A. J.
  1909 "Evidence against alcohol," *McClure's* 32 (March): 557–66.
Rosenberg, Charles E.
  1962 "The Place of George M. Beard in Nineteenth-Century Psychiatry," *Bulletin of the History of Medicine* 36: 245–59.
  1966 "Science and American Social Thought." In *Science and Society in the United States*, edited by D. D. Van Tassel and M. G. Hall, pp. 135–62. Homewood, Ill.: Dorsey.
  1974 "The Bitter Fruit: Heredity, Disease and Social Thought in Nineteenth-Century America," *Perspectives in American History* 8: 189–238.
Sadler, W. S.
  1922 *Race Decadence: An Examination of the Causes of Racial Degeneracy in the United States.* Chicago: McClurg.
Saleeby, C. W.
  1910 "Racial Poisons. II. Alcohol," *Eugenics Review* 2: 30–52. Nontransmission of "acquired characters" is not inconsistent with damage to the germ by alcohol.
  1911 *Woman and Womanhood.* New York: Mitchell Kennerley. Al-

cohol as "The Chief Enemy of Women" because of hereditary effects.
Shaw, George Bernard
  1908 *The Sanity of Art: An Exposure of the Current Nonsense about Artists being Degenerate.* London: The New Age Press, New York: Tucker. Reprinted from *Liberty* magazine (1895) with a new introduction.
Sinclair, Andrew
  1962 *Prohibition: The Era of Excess.* Boston: Little, Brown.
Stockard, Charles R.
  1913 "The effect on the offspring of intoxicating the male parent and the transmission of the defects to subsequent generation," *American Naturalist* 47: 641–82. Experiments on guinea pigs.
Stocking, George W., Jr.
  1962 "Lamarckianism in American Social Science: 1890–1915," *Journal of the History of Ideas* 23: 239–56.
Swart, K. W.
  1964 *The Sense of Decadence in Nineteenth-Century France.* The Hague: Nijhoff.
Sydow, Eckart von
  1921 *Die Kultur der Dekadenz.* 2d ed. Dresden: Sibyllen-Verlag.
Talbot, E. S.
  1898 *Degeneracy: Its Causes, Signs, and Results.* London: Scott.
Timberlake, James H.
  1963 *Prohibition and the Progressive Movement 1900–1920.* Cambridge, Mass.: Harvard University Press.
Toulouse, Édouard
  1896 *Enquête Médico-Psychologique sur les Rapports de la Supériorité Intellectuelle avec la Névropathie. I. Introduction Générale. Émile Zola.* Paris: Société d'Éditions Scientifiques.
Van Roosbroeck, G. L.
  1927 *The Legend of the Decadents.* New York: Institute des Études Francaises, Columbia University. Albert Samain as an insurgent against the tradition of scientific positivism in the arts represented by Zola, the Parnassians, Hérédia, and Leconte de Lisle.

Vuillermet, F.-A.
   1911 *Le Suicide d'une Race*. Paris: Lethielleux.
Weeks, Courtenay Charles
   1938 *Alcohol and Human Life; being partly a Revision of certain Facts and Figures in the last edition of "Alcohol and the Human Body," by the late Sir Victor Horsley and the late Dr. Mary Sturge and others*. 2d ed. London: Lewis, Chapter XI.
Wettley, Annemarie
   1959 "Entartung und Erbsünde. Der Einfluss des medizinischen Entartungsbegriffes auf den literarischen Naturalismus," *Hochland* 51: 348–58.
   1959 "Zur Problemgeschichte der 'dégénérescence,'" *Sudhoffs Archiv für Geschichte der Medizin* 43: 193–212.
Wilde, Oscar
   1891 *The Picture of Dorian Gray*. London: Ward Lock.
   1906 *The English Renaissance*. Boston and London: Luce.
Williams, Henry S.
   1908 "Alcohol and the individual." *McClure's* 31: 704–12.
Zilboorg, Gregory, and Henry, George
   1941 *A History of Medical Psychology*. New York: Norton.

Chapter VIII

Adams, Brooks
   1896 *The Law of Civilization and Decay*. New York and London: Macmillan.
Adams, Henry
   1958 "The Tendency of History" (1894); "A Letter to American Teachers of History" (1910); and "The Rule of Phase Applied to History" (1909). Reprinted with an introduction by Brooks Adams in *The Degradation of the Democratic Dogma*. New York: Putnam's, Capricorn.
Adams, J. T.
   1929 "Henry Adams and the New Physics," *Yale Review* 19: 283–302.
Allen, Garland E.
   1969 "T. H. Morgan and the emergence of a new American biology." *Quarterly Review of Biology* 44: 168–88.

Apollinaire, Guillaume
  1949 *The Cubist Painters; Aesthetic Meditations.* Translated from French by L. Abel. Rev. ed., New York: Wittenborn, Schultz (first pub. 1913).
Barber, David S.
  1968 *The Survival of the Unfittest: Evolutionary Social Thought in the Works of Henry Adams.* Ph.D. Dissertation, University of Michigan.
Barzun, Jacques
  1962 "From the Nineteenth Century to the Twentieth." In *Chapters in Western Civilization,* 3rd ed., ed. Bernard Wishy, Vol. II, pp. 340–63. New York: Columbia University Press.
  1974 "European Culture since 1800," *Encyclopedia Britannica,* 15th ed. Chicago: Encyclopedia Britannica, Inc. *Macropaedia* 6: 1066–81.
Bernfeld, S.
  1944 "Freud's earliest theories and the school of Helmholtz," *Psychoanalytic Quarterly* 13: 341–62.
  1949 "Freud's scientific beginnings," *American Imago* 6: 163–96.
Birnbaum, Lucille Terese
  1965 *Behaviorism: John Broadus Watson and American Social Thought, 1913–1933.* Ph.D. Dissertation, University of California, Berkeley.
Bukharin, N. I.
  1925 *Historical Materialism: A system of Sociology.* New York: International Publishers. Translation from 3rd Russian ed. Theory of equilibrium and phase transitions of society.
Burnham, J. C.
  1960 "Psychiatry, psychology and the progressive movement," *American Quarterly* 12: 457–65.
Chesterton, G. K.
  1942 "To Edmund Clerihew Bentley. The Dedication of the Man who was Thursday." Poems (Collected, 1915). In *The Collected Poems of G. K. Chesterton,* 9th ed. London: Methuen, pp. 109–110.
Cravens, Hamilton, and Burnham, John
  1971 "Psychology and Evolutionary Naturalism in American Thought, 1890–1940," *American Quarterly* 23: 635–57.

Davis, Douglas
  1973 *Art and the Future. A History/Prophecy of the Collaboration between Science, Technology and Art.* New York: Praeger.
De Mott, Benjamin
  1963 "Science and the rejection of realism in art," *Synthese* 15: 389–400.
Drake, Durant; Lovejoy, Arthur O.; Pratt, James Bissett; Rogers, Arthur K.; Santayana, George; Sellars, Roy Wood; Strong, C. A.
  1920 *Essays in Critical Realism: A co-operative study of the problem of knowledge.* London: Macmillan.
Einstein, Albert.
  See Laporte.
Elgar, Frank, and Maillard, Robert
  1957 *Picasso.* 2d ed. London: Thames & Hudson.
Felix, Lucienne
  1957 *L'Aspect Moderne des Mathématiques.* Paris: Blanchard, Chapters I, II: "La Revision des Valeurs au debut du XX$_e$ Siecle."
Feuer, L. S.
  1971 "The social roots of Einstein's theory of relativity," *Annals of Science* 27: 277–98, 313–44; also in his *Einstein and the Generations of Science.* New York: Basic Books, 1974, Chapter I.
Franklin, W. S.
  1910 "On Entropy," *Physical Review* 30: 766–75.
Freud, Sigmund
  1955 "Psychoanalyse und Telepathie." Ms. dated 1921. English translation in *The Complete Psychological Works of Sigmund Freud,* ed. J. Strachey. Vol. 18, pp. 177–93. London: Hogarth.
Gabo, Naum
  1965 "The Constructive Idea in Art." In *Modern Artists on Art,* ed. R. L. Herbert, pp. 104–13. Englewood Cliffs, N.J.: Prentice-Hall. First published in 1937.
Gabo, Naum, and Pevsner, Noton
  1920 *Realisticheskii Manifest.* Moscow. Reprinted with English translation in *Gabo: Constructions, Sculpture, Paintings, Drawings, Engravings,* Cambridge, Mass.: Harvard University Press (1957), pp. 151–52.

Gibbs, J. Willard
- 1876 "On the equilibrium of Heterogeneous Substances," *Transactions of the Connecticut Academy* 3: 108–248 (1876), 343–524 (1878). Reprinted in *The Collected Works of J. Willard Gibbs.* New Haven: Yale University Press (1947), Vol. I.

Glicksberg, Charles I.
- 1947 "Henry Adams and the Repudiation of Science," *Scientific Monthly* 64: 63–70.

Holt, Edwin B.; Marvin, W. T.; Montague, W. P.; Perry, R. B.; Pitkin, W. B.; and Spaulding, E. G.
- 1912 *The New Realism: Cooperative Studies in Philosophy.* New York: Macmillan.

Holton, Gerald
- 1968 "Mach, Einstein, and the Search for Reality," *Daedalus* 97: 636–73.

Illy, Jozsef
- 1971 "On the birth of Minkowski's four-dimensional world," *Actes du XIIIe Congrès International d'Histoire des Sciences, Moscow, 1971.* Moscow: Nauka (1974), Vol. 6: 67–72.

Jordy, W. H.
- 1952 *Henry Adams: Scientific Historian.* New Haven: Yale University Press.

[Kelvin] Thomson, William
- 1884 *Notes of Lectures on Molecular Dynamics and the Wave Theory of Light.* Baltimore: Johns Hopkins University.
- 1891 "Electrical Units of Measurement" (1883). In his *Popular Lectures and Addresses*, 2d ed. London: Macmillan, Vol. 1, pp. 80–143.

Keyser, Cassius Jackson
- 1947 *Mathematics as a Culture Clue and other Essays.* New York: Scripta Mathematica, Yeshiva University. Arguments supporting Spengler.

Kimball, Arthur Lalanne
- 1906 "The relations of the science of physics of matter to other branches of learning." *Congress of Arts and Science, Universal Exposition, St. Louis, 1904,* Vol. 4: 69–86. Boston: Houghton Mifflin (1906). Revival of atomism.

Krutch, J. W.
  1929 *The Modern Temper: A Study and a Confession.* New York: Harcourt, Brace.
Kuhn, Thomas S.
  1961 "The function of measurement in modern physical science," *Isis* 52: 161–93.
Laporte, Paul M.
  1966 "Cubism and Relativity, with a letter of Albert Einstein," *Art Journal* 25: 246–48. Einstein's letter denies the alleged connection.
Loeb, Jacques
  1915 "Mechanistic science and metaphysical romance," *Yale Review* 4: 766–85.
Mach, Ernst
  See Toulmin
May, Henry F.
  1959 *The End of American Innocence: A Study of the First Years of Our Own Time 1912–1917.* New York: Knopf.
McCormmach, Russell
  1974 "On academic scientists in Wilhelmian Germany," *Daedalus* 103: 157–72.
Mendelssohn, K.
  1973 *The World of Walther Nernst: The Rise and Fall of German Science 1864–1941.* Pittsburgh: University of Pittsburgh Press.
Meyerson, Emile
  1930 *Identity and Reality.* Translated from 3rd French ed. London: Allen & Unwin. The 1st French ed. appeared in 1908.
Mitchell, Donald
  1966 *The Language of Modern Music.* New York: St. Martin's Press.
Mumford, Lewis
  n.d. "Apology to Henry Adams," tape BB 0252. Los Angeles: Pacifica Foundation Tape Library.
  1951 "From Revolt to Renewal." In *The Arts in Renewal,* pp. 1–31. Philadelphia: University of Pennsylvania Press.
Nernst, Walther
  1918 *Theoretische und experimentelle Grundlagen des neuen Wärmesatzes.* Halle: Knapp. English translation from the second

German edition, *The New Heat Theorem*. London: Methuen, 1926. Chapter XIV deals with degeneration of gases.
1919 "Einige Folgerungen aus der sogenannten Entartungstheorie der Gase." *Sitzungsberichte der preussischen Akademie der Wissenschaften* (Berlin), 118–27.
See also Mendelssohn.

Nichols, Roy F.
1935 "The dynamic interpreation of history," *New England Quarterly* 8: 163–78. On Henry Adams.

Nye, Mary Jo
1975 "Science and Socialism: The Case of Jean Perrin in the Third Republic," *French Historical Studies* 9: 141–69.

O'Neil, William M.
1968 "Realism and behaviorism," *Journal of the History of Behavioral Sciences* 4: 152–60.

Perry, Ralph Barton
1912 *Present Philosophical Tendencies, A Critical Survey of Naturalism, Idealism, Pragmatism and Realism, together with a synopsis of the Philosophy of William James.* New York: Longmans, Green.

Pfaundler, Leopold
1904 *Die Physik des täglich Lebens.* Stuttgart and Leipzig: Deutsche Verlags-Unstalt.

Planck, Max
See Toulmin.

Purcell, Edward A., Jr.
1973 *The Crisis of Democratic Theory: Scientific Naturalism and the Problem of Value.* Lexington: The University Press of Kentucky.

Rauber, D. F.
1972 "Sherlock Holmes and Nero Wolfe: The role of the 'great detective' in intellectual history," *Journal of Popular Culture* 6: 483–95. "The argument is that Sherlock Holmes reflects . . . the basic assumption and tones of classical physics, while Nero Wolfe . . . exhibits marked differences which correspond . . . to the revolutionary changes in physics produced by the emergence of sub-atomic phenomena," e.g., Wolfe relies on abstract mathematics.

Richardson, John Adkins
  1971 *Modern Art and Scientific Thought.* Urbana: University of Illinois Press.
Russell, Bertrand
  1959 *My Philosophical Development.* London: George Allen & Unwin.
Sachs, Mendel
  1970 "Positivism, realism, and existentialism in Mach's influence on contemporary physics," *Philosophy and Phenomenological Research* 30: 403–20. The "Mach principle" in relativity is an example of realism, in contrast to his positivist influence as seen in the Copenhagen interpretation of quantum theory.
Samuels, Ernest
  1964 *Henry Adams: The Major Phase.* Cambridge, Mass.: Harvard University Press.
Schanck, Richard L.
  1954 *The Permanent Revolution in Science.* New York: Philosophical Library.
Schneer, Cecil J.
  1969 *Mind and Matter.* New York: Grove Press, Chapter 12, "The New Mechanism."
Schuster, Arthur
  1924 *An Introduction to the Theory of Optics.* 3rd ed. London: Arnold.
Spengler, Oswald
  1918 *Der Untergang des Abendlandes.* München: Beck.
  1962 *The Decline of the West.* English abridged edition by Arthur Helps, from the translation by C. F. Atkinson. New York: Knopf.
Taylor, F. W.
  1911 *The Principles of Scientific Management.* New York: Harper.
Toulmin, Stephen, ed.
  1970 *Physical Reality.* New York: Harper & Row. Includes the 1909–10 debate between Planck and Mach.
Wasser, Henry
  1956 *The Scientific Thought of Henry Adams.* Thessaloniki.
Welch, Robert
  1961 *The Blue Book of the John Birch Society*, 4th printing. Applica-

tion of Spengler's theory of history to American politics.
Werkmeister, W. H.
 1949 *A History of Philosophical Ideas in America*. New York: Ronald Press, Chapter 17.
Zamyatin, Yevgeny
 1970 "On Literature, Revolution, Entropy, and other Matters" (1923). English translation in *A Soviet Heretic: Essays by Yevgeny Zamyatin*, edited by M. Ginsburg, pp. 107–12. Chicago: University of Chicago Press.

Index

Ackerknecht, E. H., 141
Adams, B., 121-23, 190
Adams, C., 122
Adams, H., 121-25, 126-27, 167, 190
Adams, J., 122
Adams, J. T., 190
aestheticism, 7, 26, 91, 106
aether, *see* ether
Agassiz, L., 35
agnosticism, 23, 89
Albert, Prince, 82
Albritton, C. C., 149
alcohol, 114-20
Aliotta, A., 167
Allen, G. E., 110, 180, 190
American science, 47
Ampère, A. M., 77
Andersson, O., 180
Andler, C., 158
anthropology, 91, 125
Antoine, J.-C., 158
Apollinaire, G., 121, 191
architecture, 17, 129
Aristotle, 3, 5, 50
Arnheim, R., 167
Arnold, M., 2, 137
Artigiani, P. R., 167
Ashby, E., 141
astronomy, 30, 34, 36, 67-68, 93
atheism, 23, 89
atmosphere, 52-54
atom: atomism, 7, 11, 19, 85-86, 93; collisions, 12, 13, 66, 99; forces, 19, 21, 69; motions, 12, 20-21, 64; nucleus, 4; real existence, 12, 85-86, 99-101, 126, 133-34; size, 11, 86; weight, 11, 53

Avenarius, R., 94, 167
Ayres, C. E., 167

Badash, L., 84, 167
Balaguer Periguell, E., 180
Balzac, H. de, 23
Barber, D. S., 191
Barker, E., 167
Barzun, J., 22, 23, 25, 127, 135, 141, 180, 191
Baudelaire, C., 103, 105-106, 109, 180
Beale, O. C., 180
Beard, G. M., 104, 180-81, 188
Becker, O., 158
Beethoven, L. v., 17
behaviorism, 129
Benda, J., 167-68
Benn, A. W., 141
Benthamism, 8
Bergson, H., 14, 92, 168
Berlioz, H., 17
Bernfeld, S., 158, 191
Berthelot, M., 179
Berthelot, R., 168
Berzelius, J. J., 24
Bever, T. G., 9, 137
Bevington, M. M., 149
Biermann, K.-R., 97, 168
Binkley, R. C., 141
biology (*see also* evolution), 18-19, 22, 24, 41, 56, 88-89, 94, 134
Birnbaum, L. T., 191
birth control, 110
Bitterlich, M., 181
black body radiation, 55, 70-71
Blake, W., 17
Blanqui, A., 75n, 158
Boas, F., 56, 154

199

Bocheński, I. M., 8, 137
Bohr, N., 4, 127
Boltwood, B., 43, 149
Boltzmann, L., 78, 128; kinetic theory, 12-13, 53, 85, 95, 96, 159; radiation law, 55; statistical interpretation of 2nd law, 66-67, 69-71, 76, 98, 159, 168
Boring, E. G., 137, 168
Boscovich, R., 19, 21, 69, 73
Boulenger, M., 181
Bourget, P., 170
Boutroux, E., 168
Bowle, J., 141
Bozeman, T. D., 142
Bradley, F. H., 92
Brancusi, C., 130
Brand, L., 181
Brewster, D., 178
Brinton, C., 6, 137, 142
Brougham, H., 26
Brouwer, L. E. J., 128
Brown, A. W., 168
Brown, R., 24
Brownian movement, 98, 126
Brücke, E., 129
Brunhes, B., 159
Brush, S. G., 21, 47-48, 54, 71, 135, 137, 141, 142, 154, 159, 168
Bryan, G. H., 159
Bryan, W. J., 114-15
Buchanan, J., 168
Büttner, A., 169
Bukharin, N. I., 191
Bunke, O., 181
Burbury, S. H., 70, 71, 159-60
Burchfield, J., 52, 149-50
Burke, J., 150
Burne-Jones, E., 91
Burnham, J. C., 191
Burstyn, H., 52
Butterfield, H., 3, 137

caloric, 9-10, 12, 21, 31
Cameron, F. K., 169
Cannon, W. [=S.] F., 142
Cantor, G., 97
Čapek, M., 160
Carathéodory, C., 160

Carbonelle, I., 169
Carlyle, T., 18
Carnot, S., 9-11, 30, 127, 150
Caro, E. M., 169
Carpenter, E., 169
Carrere, J., 181
Carter, A. E., 181
Carus, P., 169
Caspari, O., 73
Cassirer, E., 169
catastrophist geology, 33, 35, 42, 57
catholicism, neo-catholicism, 17, 92, 93, 109, 110-11, 114
Charcot, J. M., 120
Chase, A., 181
chemistry, 11, 19, 24, 52-53
Chesterton, G. K., 103, 191
Chiarello, R. J., 9, 137
Clark, P., 96, 172
Clark, X., 97, 169
classicism, neoclassicism, 16-17, 126
Clausius, R., 13, 21, 31, 53, 65, 77, 124, 127, 150, 160
Clifford, W. K., 104, 181
Clive, J., 142
Cocke, W. J., 160
Coleridge, S. T., 17, 20
collectivism, 8
Collingwood, R. G., 137-38
colors, 19
Commager, H. S., 142
Comte, A., 23, 55-56, 59, 93, 154, 169
conservation: force or energy, 10, 14, 20, 63, 95; mass-energy, 128
continental drift, 48, 58-59
Cope, E. D., 42, 150
Copenhagen interpretation of quantum mechanics, 127
Copland, A., 23, 107, 142, 181
cosmography, 56-57
Courbet, G., 23
Cournot, A. A., 97, 169
Cowan, R. S., 181
Crafts, W. F., 182
Crane, H. R., 49, 154-55
Cranefield, P. F., 143

INDEX

Cravens, H., 169, 191
Creasey, C. H., 170
Croce, B., 91
Croll, J., 97, 169
Crosland, M. P., 53, 155
Crowe, M., 5, 138
cubism, 126-27, 130, 131
Culotta, C. A., 143
culture (definition), 2
Cuneo, E., 138
Curie, M., 42
Curie, P., 42
cyclic history (see also eternal return, recurrence), 6, 8, 64, 71, 72

dada, 126
Dalton, J., 47, 52-53
Damm, A., 182
Darwin, C., 6, 13, 23, 34, 36, 38-42, 77-78, 123, 149, 150
Darwin, E., 150
Darwin, G. H., 57, 155
Daub, E., 160
Daumier, H., 23
Dauvillier, A., 160
Davenport, C. B., 111-12, 182
Davenport, G. C., 111
Davie, G. E., 143
Davis, D., 192
Davy, H., 20, 21, 24, 143, 144
death instinct, 65
Debussy, A. C., 107
decadence, decadents, 7, 105, 109
degeneration, 2, 13-14, 26, 42, 103-20, 122, 125, 129, 134; in physics, 128
Delevsky, J., 160
De Mott, B., 192
De Musset, A., 17
denudation, 38, 41
determinism, 13, 71, 74, 90, 97-98
Dewey, J., 127
Dicey, A. V., 8, 138
Dickens, C., 23
Diderot, D., 17
Diederichs, E., 91
Dingle, H., 170
Dirac, P. A. M., 4

disorder, see randomness
dissipation of energy, 2, 10, 13, 14, 27, 29-31, 62, 74, 96, 123-25
dissolution, 62-64
Dixon, E. T., 174
Dodd, G., 143
Dollo, L., 161
Doran, R. E., 182
Dostoevsky, F. M., 23
Drake, D., 192
Drake, S., 170
Du Bois-Reymond, E., 24, 89, 96, 170
Duchamp, M., 130
Duhem, P., 92, 95, 170
Dulong, P. L., 54
Duncan, D., 138, 160
Durkheim, É., 8, 91
Dyson, F. J., 48, 155

earth: age, 34-36, 41-43; cooling, 29, 30-37, 43; rigidity, 57-58; rotation, 36-37
economics, 33
Eddington, A. S., 61, 160
efficiency, 11
Ehrenfest, P., 155, 161
Ehrenfest, T., 161
Einstein, A., 4, 34, 126, 127, 128, 130-32, 194
Eiseley, L., 41, 150
Eisen, S., 170
elasticity, 99-101
Elderton, E. M., 116-18, 120, 182, 187
electromagnetism, 20, 24, 70-71, 77, 131
Elgar, F., 130, 192
Eliade, M., 161
Elkana, Y., 10, 96, 138, 170
Ellegård, A., 143
Elwood, E. S., 182
empiricism, empirio-criticism, 7, 26, 86, 94-95, 128, 130-31
energetics, 7, 26, 68, 73, 95-96, 122
energy (see also dissipation), 10, 20, 78, 95, 123
Engels, F., 23, 170
English literature, 17, 106

Enlightenment, 7, 14, 26
Enriques, F., 170
Ensch, N., 181
entropy, 13, 30-31, 61, 66-70, 123-26, 127; origin of word, 31
environment, 48-49
Epicurus, 98
Eriksson, G., 143
eternal return (*see also* recurrence), 14, 72-74, 123
ether, 10, 85, 123, 131
Euclid, 3
eugenics, 110-14, 117
Euler, L., 3, 4, 52
Eve, A. S., 43, 150, 170
evolution, 14, 62-65; Darwinian, 6, 13, 14, 23, 36, 40-42, 68-69, 77-78, 88-89, 115-16, 124-25; of science, 5; retrograde, *see* degeneration
expressionism, 130

Faraday, M., 20, 24, 77
fascism, 22
Fechner, G., 24, 65, 91, 161
Feitelberg, S., 158
Felix, L., 192
Féré, C. S., 182
Ferguson, D. N., 107, 182
Fermi, E., 48
Feuer, L. S., 192
Feuerbach, L. A., 23
Feuillerat, A., 170
Fichte, J. G., 16, 18
fin-de-siècle, 107-108
Fink, A. E., 182
Fischer-Homberger, E., 182
Fiske, J., 161
FitzGerald, G. F., 96, 171
Flammarion, C., 161
Flaubert, G., 23
Flugel, J. C., 161
Flugel, O., 171
Foote, G. A., 143
Forbes, J., 9
force, 10, 19-21, 100-101
Foster, M. P., 182
Fouillée, A. J. E., 171
Fourier, J. B. J.; heat conduction theory, 9, 21, 31-32, 35, 52, 150-51, 155; terrestrial temperatures, 35, 52, 155
Frank, P., 86, 171
Franklin, W. S., 128, 192
Freeman, D., 91, 171
French literature, 17, 23, 103-106
Fresnel, A., 9
Freud, S., 65, 120, 127, 129, 130, 161, 183, 192
Friedlander, R., 183
Frobenius, G. F., 97
Fullmer, J. Z., 143
functionalism, 132
futurism, 129

Gabo, N., 130, 192
Galaty, D. H., 143
Galdston, I., 143
Galilei, G., 52, 128
Galippe, V., 183
Galois, E., 22
Galton, F., 24, 80, 82, 110, 121, 171, 179, 183
Garber, E., 52, 155
gardens, 15
Gardiner, P., 138
Garland, H., 171
gases, *see* kinetic theory
Gasman, D., 92, 171, 183
Gauss, C. F., 47
Gautier, T., 105
Gay-Lussac, J. L., 53
Geikie, A., 38-39, 41, 58, 151, 155
genetics, 94, 113, 126
Genil-Perrin, G.-P.-H., 183
geography, 56
geology, 14, 31, 33-42, 45, 51, 52, 57-60, 125
geometry, 22, 24, 121, 131
geophysics, 49, 51
German literature, 16
gestalt, 8, 26, 91
Gibbs, J. W., 96, 123, 193
Gillispie, C. C., 135, 151
Gillmor, C. S., 155
Glass, B., 94, 143, 171
Glicksberg, C. I., 193
God, 15, 30, 79-82

Gode-von Aesch, A. G. F., 144
Goethe, J. W.˙ v., 16, 18-19, 109, 129
Gogol, N. V., 23
Gold, M., 183
golden age, 3-4, 14, 73
Goodfield, J., 154
Goodstein, J. R., 144
gothic revival, 17
Gottschalk, L., 9, 138
Gould, S. J., 161
Gower, B., 144
Goya, F. J., 23
Grabo, C., 144
Grant, M., 104, 183
Greenough, G., 29, 151
Griffiths, R., 171
Gropius, W., 127
group theory, 22, 97
Guerlac, H. E., 172
Gustafson, A., 183

H curve, 69
H theorem, 13, 66
Haber, F., 151
Haeckel, E., 42
Hahn, R., 172
Haines, G., IV, 172
Hall, A. R., 6
Hall, E. W., 144
Haller, J. S., Jr., 104, 183
Haller, M., 184
Hallier, E., 135
Hamilton, W. R., 21
Hardy, T., 23
Harriman, Mrs. E. H., 111
Harris. F., 161
Hartmann, E. v., 161
Hayes, C. J. H., 172
heat, 2, 9-13, 21, 95, 133; conduction, *see* Fourier; radiant, 9; transfer laws, 54–55
heat death, 31, 54, 61-76
Hegel, G. W. F., 8, 16
Heimann, P. M., 144, 161-62
Heine, H., 16, 73
Heisenberg, W., 4
Hellpach, W., 184
Helm, G., 95

Helmholtz, H. v., 24, 31, 47, 65, 127, 151
Hennemann, G., 144
Henry, G., 190
Herapath, J., 21, 53, 155-56
heredity, 94, 104-105, 110, 116, 121-22
Hermann, A., 172
Herschel, J., 40
Herschel, W., 9
Hesse, M. B., 144
Hexter, J., 138
Hibben, J. G., 172
Hiebert, E., 162, 172
Hilbert, D., 128
Hildebrandt, K., 184
Hill, C., 9, 138
Himstedt, F., 42, 151
Hirsch, W., 184
historicism, 8
historiography: contextual, 4; cyclic, 72; horizontal, 2, 4-5, 15; Kuhnian, 5; materialistic, 23; neo-idealist, 91 romantic, 17; thermodynamic, 123-24; Tory, 3-4; vertical, 2-4
Hitler, A., 113
Hobbes, T., 8
Hobson, R. P., 118-19, 184
Höffding, H., 144
Hofmannsthal, H. v., 177
holism, 8, 9
Hollingdale, R. J., 162
Holmes, S., 195
Holmes, S. J., 184
Holt, E. B., 193
Holton, G., 5, 135, 138-39, 145, 193
Hook, S., 145
Hooker, J. D., 39
Hopkins, W., 57, 156
Horsley, V., 184, 190
Howe, S. G., 116, 184
Howson, C., 172
Hoyle, F., 88, 172
Hughes, H. S., 139, 172
Hugo, V., 17, 22
humanism, 3, 91
Humboldt, A. v., 56

Hutton, J., 14, 29, 33
Huxley, T. H., 151, 172-73; Darwinism, 13, 41, 68-69; geology, 37, 42; Kelvin, 37-38, 41, 44; materialism, 89-90
Hyatt, A., 42
hysteria, 108-9

Ibsen, H., 109
ice age, 35
idealism, 7, 8, 25, 86-87, 92
Iggers, G. G., 139
Illy, J., 193
immigration restriction, 110, 113-14
impressionism, 7, 26, 91
indeterminism, *see* randomness
individualism, 8, 17, 109
intelligence, 110, 129
irreversibility, 11, 30-32, 52, 68, 70-71, 75, 125, 128

Jäger, G., 96
Jaki, S. L., 145, 162, 173
Jammer, M., 145, 173
Jeans, J., 61, 162
Jeffreys, H., 156
Jellett, J. H., 90, 173
Jensen, J. V., 173
Jodl, F., 173
Johnson, A., 114
Johnson, R. H., 117, 188
Jones, B. C., 184
Jordan, D. S., 184
Jordy, W. H., 193
Josephson, M., 185
Joule, J. P., 21, 78, 127
Joyce, C. R. B., 83-84, 173
Joyce, J., 127

Kaiser, W., 172
Kampf, A., 139
Kant, I., 8, 19-20
Kapp, R. O., 162
Kargon, R., 21, 145
Kaufmann, W., 162
Kelland, P., 32, 151
Kelvin, Lord (William Thomson), 32, 47, 68, 73, 77, 123, 193; age of earth, 14, 34-44, 57, 121, 125, 152; atoms, 11, 86, 139, 173; dissipation of energy, 30-31, 65, 124, 152, 162; heat conduction theory, 32; mechanical models, 24, 44, 133; planetary science, 55, 57, 59, 156; quantification, 132; randomness, 70; reversibility paradox, 66
Kende, M., 185
Kennedy, E. S., 163
Kern, S., 185
Keyser, C. J., 193
Kierkegaard, S., 23
Kimball, A. L., 193
kinetic theory of gases, 12-13, 21, 24, 53-54, 65-70, 75, 77, 88, 96-101
King, C., 41-42, 121, 122, 152
Klein, M. J., 163, 173
Kleinpeter, H., 173
Knight, D., 145
Knopoff, L., 51, 156
Knott, C. G., 58, 156, 163
Knowles, J. T., 139
Koch, R., 24
Koren, J., 185
Krauss, F., 185
Krönig, A., 21
Kronecker, L., 97
Krutch, J. W., 194
Kuhn, T. S., 5, 6, 83, 139, 145, 194
Kuhn, W., 173

Lagrange, J. L., 3, 52, 68
Lakatos, I., 96, 172
Lalande, A., 163
Lamarck, J. B. P. A. de M., 42, 104, 119, 121
Lange, F., 185
Lange, F. A., 174
Laplace, P. S. de, 3, 9, 20, 21, 52, 53, 68, 97, 174
Laporte, P. M., 194
latent heat, 9
Lavoisier, A. L., 9
Leavis, F. R., 2, 139
Lebon, F., 176
Le Conte, J., 152-53

Index

Léger, F., 130
Legrain, P.-M., 185, 186
Leibbrand, W., 185
Lenin, V. I., 174
Leonard, N., 145
Leoncavallo, R., 23
Leppmann, A., 185
Leppmann, F., 185
Leverette, W. E., Jr., 145
LeVerrier, U. J. J., 33-34
Liebenow, C. H., 42, 153
light, theories of, 9-10, 77, 81, 83, 133
Lilley, S., 146
Lindbergh, C., 87, 174
Lindsay, R. B., 146
Lipman, T. O., 22, 146
literature, see French, etc.
Littledale, R. F., 79-80, 174
Littré, E., 92, 174
Lodge, O., 97, 174
Loeb, J., 126, 194
Lowith, K., 163
Lombroso, C., 103, 185
Lorentz, H. A., 96
Loschmidt, J., 54, 66, 68, 86, 163, 174
Lovejoy, A. O., 15-16, 146, 192
Lucas, P., 105, 185
Ludmerer, K. M., 186
Ludwig, K. F. W., 24
Lydston, G. F., 104, 186
Lyell, C., 14, 33, 57, 121, 124, 153

MacDonald, A., 105, 186
MacDowell, E., 145
Mach, E., 70, 94, 99, 130, 163, 175, 196
Mackin, J. H., 156
MacLeod, R. M., 175
Maeterlinck, M., 109
Magendie, F., 148
Magnan, V., 105, 186
Mahler, G., 107
Maillard, R., 130, 192
Malthus, T. R., 8, 110
Mandelbaum, M., 135
Manley, G., 53, 156
Manuel, F., 139

Marchant, J., 153
Martindale, C., 9, 139, 186
Martineau, H., 186
Marvin, W. T., 193
Marx, K., 23
Mascagni, P., 23
Mason, R. O., 186
Mason, S., 186
materialism, 1, 7, 12, 14, 20, 23, 24, 73-74, 78, 85-90, 92, 93, 122, 129
mathematics, 18, 21-22, 24, 39, 86-87, 97, 128, 132-34
Matson, F. W., 175
Maxwell, J. C.: demon, 65-66, 75, 101, 122, 163; electromagnetic theory, 24, 77, 131, 133; kinetic theory, 12-13, 21, 53; Saturn's rings, 54, 55, 156; Spencer, 1, 64
May, H. F., 194
Mayer, [J.] R., 20, 73, 95, 127, 146
M'Cosh, J., 81-82, 175
McCormmach, R., 194
McKim, W. D., 186
McTaggart, J. M. E., 92
Mead, G. H., 146
Means, J. O., 175
mechanics, 12, 13, 14, 89, 125
mechanism, mechanistic theories, 1, 5, 7, 9, 12, 14, 23, 24, 68, 70, 75, 85-86, 94-95, 122, 128-29
medicine, 18, 78-79, 108, 115-16
Melloni, M., 9
Menard, H., 60, 156-57
Mencher, S., 146
Mendel, G., 94, 120, 126
Mendelsohn, E., 22, 146, 194
Mendoza, E., 153
Mercury (planet), advance of perihelion, 34
Merz, J. T., 135
meteorology, 50, 52-53
methodology of scientific research programmes, 96-97
Meyer, D. H., 175
Meyerson, E., 194
Michelson, A. A., 84-85, 175
Michelson-Morley experiment, 131

Middle Ages, 11n, 16-17
Mill, J. S., 19, 23, 93
Miller, E. C. L., 186
Millikan, R. A., 84, 176
Milne, E. A., 163
Milner, G., 186
Minkowski, H., 131
Mitchell, D., 194
modernism, 129
Möbius, P. J., 186
molecules, *see* atom
Momigliano, A. D., 72, 163
Mondrian, P., 130
money, 33, 38
Montague, W. P., 193
Mora, G., 139
Morel, B. A., 103-104, 116, 187
Morgan, C. L., 174
Morris, W., 91
Mosse, G. L., 92, 176, 187
Mott, F. W., 187
Mott, N. F., 176
Mumford, L., 130, 194
music, 16-17, 23, 107, 126, 131; perception, 9
Musset, A. de, 17
Mussorgsky, M., 23
mysticism, 7, 11, 81, 92, 108

Nägeli, K. W. v., 94
Nation, C. A., 119, 187
naturalism, 7, 23, 109, 132
Naturphilosophie, 7, 18, 22, 96
Nazism, national socialism, 22, 92, 113
neolamarckism, *see* Lamarck
neorealism, 7, 126-32
neoromanticism, 7, 14, 26, 29, 44, 68, 86, 90-103, 107-108, 129; origin of word, 91
Nernst, W., 127, 194-95
neurasthenia, 104
Newton, I., 3-4, 52; cooling law, 54; mechanics, 12, 34, 66, 125; optics, 19, 81; world view, 6, 85, 128
Nichol, J. P., 32
Nichols, R. F., 195
Nietzsche, F., 72-74, 76, 109, 122, 164

Nordau, A., 187
Nordau, M., 107, 122, 187
Nordenskiöld, E., 146
nudity, 95, 128-29
Nye, M. J., 176, 195

Oersted, H. C., 20, 77, 146
Oken, L., 27
Oldham, R. D., 58, 157
Olson, R., 135
O'Neil, W. M., 195
Opper, J., 136, 146-47
organicism, 12, 23
Ostwald, W., 25, 95, 96, 122, 147
Oxford movement, 17, 109

Packard, A., 42
Packard, V., 176
painting, 16, 23, 121, 130
paradigm, 5
Parrington, V. L., 176
Parsons, T., 8, 139, 176
Pasteur, L., 24
patriotism, 17
Paul, H. W., 176
Pearson, K., 110, 111, 116-18, 120, 176, 182, 187
Peirce, C. S., 97, 176-77
Penrose, L. S., 164
Periodicals, 26-27
Perrin, J., 126
Perry, R. B., 129, 193, 195
Persons, S., 147
Pessimism, 2, 29-30, 72-73, 96, 107-108, 123
Peterson, H., 69, 93, 164, 177
Petit, A. T., 54
Petrazzani, P., 187
Pevsner, N., 192
Pfaundler, L., 128, 195
Pfeffer, R., 164
Pfeifer, E. J., 153
Phase rule, 123-24
Phillips, D. C., 177
philosophy, 23, 25, 129
physics (*see also* heat), 4, 8, 14, 19-20, 39, 41, 46-52, 59, 77, 84; classical (Newtonian) 12; "physicists" 79-80; theoretical 1, 4, 21, 24, 67

Physics Survey Committee, 50-51, 157
Picasso, P., 127, 130, 131
Pickens, D. K., 187
Pickett, D., 188
Pitkin, W. B., 193
Planck, M., 4, 55, 59, 71-72, 85, 126, 157, 164
planetary science, 46-60
Plato, 110, 130
Playfair, J., 36, 39, 153
poetry, 9, 16
Poincaré, H., 68, 72, 75-76, 95, 97, 164, 177
Poisson, S. D., 21, 68
politics, 8, 17, 23, 91, 115
Popenoe, P., 112-13, 117, 188
Popper, K., 140
positivism, 1, 23, 86, 92-94; logical, 101
Potts, W. A., 188
Pratt, J. B., 192
prayer test, 78-84, 134
Priestley, J. B., 136
Primer, S., 177
Pritchard, C., 27, 147
probability, 13, 67, 125
progress, 3
progressive movement, 110-12, 114, 126, 129
prohibition (of alcohol), 114-19
Protestant revival, 17
psychoanalysis, 91, 120, 128-29
psychology, 91, 129
psychophysics, 24, 91
Ptolemy, 3
Purcell, E. A., Jr., 195
pure science, 46-47
purpose, 12

qualitative issues in science, 133
quantification, 78, 132-34
quantum theory, 4, 12, 55, 59, 85, 87, 126, 127, 128
Quetelet, A., 24

racism, 92, 104-105, 112, 113, 115
radiant heat, 9, 55
radicalism (philosophical), 8

radioactivity, 42
radium, 42-43
Rádl, E., 147
Rainoff, T. J., 140
Rand, W., 2-3, 140
randomness, 13, 30, 40, 67, 70-71, 97-99, 125
Rankine, W. J. M., 65, 164
rationalism, 7, 23
Rauber, D. F., 195
Raven, C. E., 188
realism, 1, 2, 7, 8, 9, 14, 15, 18, 19, 23-26, 68, 86, 90, 109, 130
recurrence, 17, 61, 72-74; paradox 67-68, 73-76
reductionism, 8, 95
Reichenbach, H., 70, 164
Reid, G. A., 188
Reingold, N., 52
relativity, 87, 128, 130-32
religion, 17, 23, 62, 78-83, 92
Rentoul, R. R., 188
reversibility paradox, 66-67
revolution, political, 5, 6, 127
revolution, scientific, 5-6
Rey, A., 164, 177
Rice, C. S., 122
Richardson, J. A., 195
Riemann, G. F. B., 24
Riley, W., 147
Ringer, F. K., 92, 177
Rodin, A., 122
Rogers, A. K., 192
Romanell, P., 177
romanticism, 1, 2, 7, 8, 15-22, 25-27, 109, 129; death, 22, 107; origin of term, 16
Roosevelt, T., 87, 122, 177
Rosanoff, A. J., 188
Rosanoff, M. A., 188
Rosenberg, C. E., 188
Rosicrucians, 109
Rossetti, D. G., 91
Rousseau, G. S., 140
Royce, J., 92
Rudwick, M. J. S., 33, 153
Ruskin, J., 109
Russell, B., 92, 127, 196
Rutherford, E., 43-44, 57

Sachs, M., 196
Sadler, W. S., 188
Sageret, J., 177
Saleeby, C. W., 188-89
Salisbury, Lord, 69
Samain, A., 189
Samuels, E., 196
Santayana, G., 192
Saturn's rings, 54, 55
Schanck, R. L., 196
Schapiro, M., 140
Schelling, F. W. J. v., 16, 18, 19, 20
Schiller, J. C. F. v., 16
Schlegel, F. v., 16
Schleiden, M. J., 24
Schmidt, H., 165
Schnabel, F., 147
Schneer, C. J., 140, 196
Schneider, H. W., 136
Schneider, I., 53, 157
Schnitzler, A., 177
Schoenberg, A., 127, 131
Schofield, R., 8-9, 140
Schopenhauer, A., 96
Schorske, C. E., 177
Schrödinger, E., 4, 165
Schubert, F. P., 17
Schulman, L. S., 165
Schuster, A., 84, 128, 178, 196
Schwann, T., 24
Schweigger, J. S. C., 147
scientism, 93
Scott, W. L., 178
Scrope, P., 153
seismology, 58
Sellars, R. W., 192
sensationalism, 7, 29
set theory, 97
sex differences, 19
Shaffer, E. S., 147
Shakespeare, W., 16
Shaler, N. S., 45, 157
Sharlin, H. I., 153
Shaw, G. B., 110, 127, 189
Shelley, P. B., 71-72, 165
Sherlock, T. T., 174
Shryock, R. H., 18, 147
Siegfried, R., 147

Simon, W. M., 178
Sinclair, A., 189
sludge, activated, 48
Smoluchowski, M. v., 96
Smyth, W., 165
Snelders, H. A. M., 21, 147-48
Snow, C. P., 2
social science, sociology, 91, 132
Somervell, D. C., 8, 140
Sorel, G., 92
Sorokin, P. A., 140, 165, 178
space science, 48, 49-50
Spaulding, E. G., 193
Spencer, H., 23, 62-65, 165
Spencer, J. B., 148
Spengler, O., 125-26, 127, 196
stability, 14, 65, 67-68
Stallo, J. B., 99-101, 123, 178
Stambaugh, J., 165
statistical explanation, 12-13, 24, 71, 98, 117, 125-26
Stauffer, R. C., 20, 148
steam engines, 9, 30
Stebbing, S., 86, 178
Stefan, J., 55
sterilization, 112-13
Stern, F., 178
Stern, W., 129
Stewart, B., 66, 165
Stockard, C. R., 189
Strauss, D., 141
Strauss, R., 106-107
Stravinsky, I., 127
Strong, C. A., 192
Strutt, R. J., 43, 153
Sturge, M. D., 184, 190
sun, 33-34, 36, 54, 62
Suppe, F., 140
Swart, K. W., 189
Swinburne, A. C., 91, 109
Swinton, W. E., 157
Sydow, E. v., 189
symbolism, 7, 91
Sypher, W., 136, 165
Szilard, L., 166

Tait, P. G., 32, 39, 66, 153-54, 165
Talbot, E. S., 189
Tarde, G., 166

Taton, R., 136
Taylor, F. W., 129, 196
Temkin, O., 148
temperance movement, 114-16
temperature, 10, 123; variation in atmosphere, 53-54
Temple, G., 148
Terletskii, Ya. P., 166
Terman, L. M., 129
Thackray, A., 53, 157
themata, 5
thermodynamics, 24, 122-23, 127-28; 1st law, 10, 29-31, 127; 2d law (*see also* dissipation), 11, 13-14, 29, 30-31, 54, 61-62, 65, 66-67, 96, 124-25, 127; 3rd law, 127-28
Thiele, J., 178
Thomas, R. H., 148
Thompson, H., 78-79, 178
Thompson, S. P., 32, 154
Thomson, D., 148, 178
Thomson, J. J., 43, 154
Thomson, W., *see* Kelvin
Timberlake, J. H., 189
time: analogy with money, 33, 38; direction, 11, 13, 68, 70, 76; relation to space, 131-32
Tindall, W. Y., 148
Tolman, R. C., 166
Tolstoi, L., 109
Tonreihe, tone row, 126, 131
toryism (political), 8
Toulmin, S., 5, 6, 140, 154, 196
Toulouse, É., 189
transcendentalism, 17
Truesdell, C., 3-4, 141
Tuchman, B., 85, 178
Turner, F., 179
Tyndall, J., 61, 64, 78-81, 89-90, 166, 179

uniformitarian geology, 14, 33, 35, 37, 41, 57, 68, 121, 125
Unitarian Church, 17, 23
unity of natural forces, 1, 10, 20, 96, 134
utilitarianism, 8, 18

van der Waals, *see* Waals
van Roosbroeck, G. L., 189
verismo, 23
veritism, 171
Verlaine, P., 109
Victoria, Queen, 82
Virtanen, R., 179
vitalism, 18, 22, 129
Vogt, J. G., 73, 74n, 166
von Weizsäcker, C. F., 166

Waals, J. D. van der, 77, 179
Wachsmuth, B., 148
Waerden, B. L. van der, 166
Wagner, R., 23, 109
Wallace, A. R., 23, 40
Walpole, H., 17
Ward, J., 179
Wasser, H., 196
Waterston, J. J., 33, 53, 154, 157
Watson, J. B., 129, 130
wave theory: heat, 10; light, 9-10, 26
Weber, E. H., 24
Weber, M., 8, 91, 157-58
Weeks, C. C., 190
Wegener, A., 58-59, 158
Weismann, A., 116, 119
Weiss, P. A., 148
Welch, R., 196
Welldon, R. M. C., 83-84, 173
Welsh, A., 179
Werkmeister, W. H., 197
Wetterham, D., 174
Wettley, A., 185, 190
Wetzels, W. D., 149
White, L., 166
Whitman, W., 23
Whyte, L. L., 149, 179
Wiener, N., 22, 149
Wilberforce, S., 13, 68
Wilde, O., 15, 91, 106, 109, 149, 179, 190
Wilhelm, Kaiser, 122
Wilkins, T., 154
Williams, H. S., 190
Williams, L. P., 20, 147, 149

Williams, W. M., 11, 141
Wilson, D. B., 154
Wilson, F. L., 48-49, 158
Wilson, J. M., 179
Wilson, L. G., 57, 158
Wöhler, F., 22
Wolf, A., 154
Wolfe, N., 195
women's rights, feminism, 19, 23, 126
Wordsworth, W., 17
Wright, F. L., 127

Youmans, E. L., 145
Young, J., 180
Young, T., 26

Zagorin, P., 141
Zamyatin, Y., 127, 197
Zanstra, H., 166
Zawirski, Z., 166
Zermelo, E., 76, 97, 166-67
Zilboorg, G., 190
Zoellner, J., 73, 100, 180
Zola, E., 23, 105, 109, 189

Randall Library - UNCW
QC7 .B78 1978   NXWW
Brush / The temperature of history : phases of sci
3049002396654